£20·00

Finite elements for electrical engineers

Finite elements
for electrical
engineers

P.P.SILVESTER

R.L.FERRARI

CAMBRIDGE UNIVERSITY PRESS
Cambridge
London   New York   New Rochelle
Melbourne   Sydney

Published by the Press Syndicate of the University of Cambridge
The Pitt Building, Trumpington Street, Cambridge CB2 1RP
32 East 57th Street, New York, NY 10022, USA
296 Beaconsfield Parade, Middle Park, Melbourne 3206, Australia

First published 1983

Printed in Great Britain at the University Press, Cambridge

Library of Congress catalogue card number: 83-23550

British Library cataloguing in publication data
Silvester, P. P.
Finite elements for electrical engineers.
1. Finite element method
I. Title II. Ferrari, R. L.
515.3′53   TA347.F5

ISBN 0 521 25321 7 hard covers
ISBN 0 521 27310 2 paperback

JWA

Contents

Preface

Although there are now many research papers in the literature that describe the application of finite element methods to problems of electromagnetics, no textbook has appeared to date in this area. This is surprising, since the first papers on finite element solution of electrical engineering problems appeared in 1968, about the same time as the first textbook on finite element applications in civil engineering.

The authors have both taught courses in finite elements to a variety of electrical engineering audiences, and have sorely felt the lack of a suitable background book. The present work is intended to be suitable for advanced undergraduate students, as well as for engineers in professional practice. It is undemanding mathematically, and stresses applications in established areas, rather than attempting to delineate the research forefront.

Both authors wish to thank the many people who have helped shape this book – especially their students.

PPS and RLF
Montreal and Cambridge
June 1983

1

First-order triangular elements for plane problems

1. Introduction

First-order triangular finite elements made their initial appearance in electrical engineering applications in 1968. They were then used for the solution of comparatively simple waveguide problems, but have since then been employed in many areas where two-dimensional scalar potentials or wave functions need to be determined. Because of their relatively low accuracy, first-order elements have been supplanted in many applications by elements of higher orders. However, they continue to find use in problems where material nonlinearities or complicated geometric shapes are encountered; for example, in analysing the magnetic fields of electric machines, or the charge and current distributions in semiconductor devices.

The first-order methods using triangular elements may be regarded as two-dimensional generalisations of piecewise-linear approximation techniques, tools widely used in virtually all areas of electrical engineering. Furthermore, the mathematics required in defining the elements verges on the trivial, and computer programming at a very simple level can produce many useful results. There are few methods in electromagnetic field theory for which such sweeping claims can be made, and indeed it is surprising that finite elements have not penetrated into electrical engineering applications even more rapidly.

In this chapter, simple, triangular, finite element methods will be developed for solving two-dimensional scalar potential and wave problems. The construction of these simple elements is useful in its own right; but perhaps more importantly, it will also illustrate, by way of example, many of the principles involved in all finite element methods.

2.   Laplace's equation

Numerous problems in electrical engineering require a solution of Laplace's equation in two dimensions. For example, determination of the TEM wave properties of a coaxial transmission line composed of rectangular conductors, as in Fig. 1.1(a), requires finding the electric potential distribution in the interconductor space. Only one-quarter of the actual problem region needs to be analysed because of symmetry. Thus, there arise two kinds of boundary conditions in this case: prescribed potential values along the conductive metal surfaces (Dirichlet conditions), and vanishing normal derivative values along the symmetry planes. Subject to these conditions, the potential itself is governed by Laplace's equation,

$$\nabla^2 u = 0. \tag{2.01}$$

Similarly, classical analysis of electric machines requires determination of the magnetic scalar potential distribution in the air-gap region, as in Fig. 1.1(b). Again, Laplace's equation holds in the interior of this region. The boundary conditions are much like those of the electric potential problem: the scalar potential has fixed values along iron surfaces, and must have a vanishing normal derivative at symmetry planes.

The well-known principle of minimum potential energy requires that the potential distribution in the transmission line or the slot must be such as to minimise the stored field energy per unit length. To within a constant multiplier this energy is given by

$$W(u) = \tfrac{1}{2} \int |\nabla u|^2 \, \mathrm{d}S, \tag{2.02}$$

the integration being carried out over the two-dimensional problem region. This minimum-energy principle is mathematically equivalent to Laplace's equation in the sense that a potential distribution which satisfies the latter equation will also minimise the energy, and vice versa. Hence, two alternative practical approaches exist for solving the field problem. On the one hand, an approximate solution to Laplace's equation may be sought directly – as it is, for example, in the separation of variables technique, or in finite difference methods. Alternatively, an approximate expression may be created for the stored energy $W(u)$ associated with the potential $u(x, y)$ by assuming the potential u to be given by a combination of suitably chosen, simple functions with as yet undetermined coefficients. Minimisation of the energy then determines the coefficients, and thereby implicitly determines an approximation to the potential distribution. Virtually all finite element methods follow the second route, or adaptations of it.

Fig. 1.1. (*a*) One-quarter of the rectangular coaxial line, showing boundary conditions of the problem. (*b*) Half of one tooth pitch of an electric machine, and its finite element model. The heavy lines indicate Dirichlet boundaries (potential specified), whereas the remaining boundaries are planes of symmetry, requiring vanishing normal derivative of potential.

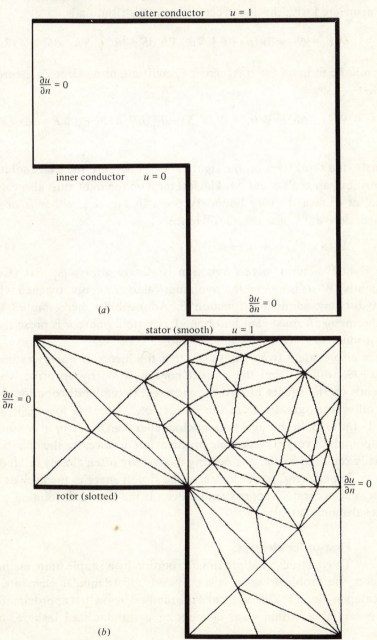

Suppose $u(x, y)$ is the true solution of the problem, while $h(x, y)$ is some sufficiently differentiable function with exactly zero value at every boundary point where the value of u is prescribed by the boundary conditions. The combination $(u + \theta h)$, where θ is a scalar parameter, then has the same prescribed boundary values as u. The energy $W(u + \theta h)$ associated with this incorrect potential distribution is

$$W(u + \theta h) = W(u) + \theta \int \nabla u \cdot \nabla h \, dS + \tfrac{1}{2}\theta^2 \int |\nabla h|^2 \, dS. \qquad (2.03)$$

The middle term on the right may be rewritten, using Green's theorem, so that

$$W(u + \theta h) = W(u) + \theta^2 W(h) - \theta \int h \nabla^2 u \, dS + \theta \oint h \frac{\partial u}{\partial n} \, ds. \qquad (2.04)$$

Clearly, the third term on the right vanishes because the exact solution u satisfies Laplace's equation. The last term on the right must also vanish since, at each and every boundary point in Fig. 1.1, either h or the normal derivative of u vanishes. Hence,

$$W(u + \theta h) = W(u) + \theta^2 W(h). \qquad (2.05)$$

The rightmost term in this equation is clearly always positive. Consequently, $W(u)$ is indeed the minimum value of energy, reached when $\theta = 0$ for any admissible function h. Admissibility here implies two requirements: h must vanish at boundary points where u is prescribed; and h must be at least once differentiable.

It is also evident from Eq. (2.05) that the incorrect energy estimate $W(u + \theta h)$ differs from the correct energy $W(u)$ by an error which depends on the square of θ. If the incorrect potential distribution does not differ very greatly from the correct one – that is to say, if θ is small – the error in energy is thus much smaller than the error in potential. This point is of very considerable practical importance, for the quantities actually required by the engineering analyst are often closely related to energy. Impedances, power losses, or the stored energies themselves are often very accurately approximated even if the potential solution contains substantial errors.

3.   First-order elements

To construct an approximate solution by a simple finite element method, the problem region is subdivided into triangular elements, as indicated in Fig. 1.1. The essence of the method lies in first approximating the potential u within each element in a standardised fashion, and thereafter interrelating the potential distributions in the various elements so as to constrain the potential to be continuous across interelement boundaries.

To begin, the potential approximation and the energy associated with it will be built up in this section. Within a typical triangular element, illustrated in Fig. 1.2, it will be assumed that the potential is adequately approximated by the expression

$$U = a + bx + cy. \tag{3.01}$$

The true solution is thus replaced by a piecewise-planar function; the smoothly-curved, actual potential distribution over the x–y-plane is replaced by a jewel-faceted approximation. It should be noted, however, that the potential along any triangle edge is the linear interpolate between its two vertex values, so that if two triangles share the same vertices, the potential will be continuous across the interelement boundary. There are no gaps in the surface $U(x, y)$ which approximates the true solution over the x–y-plane; the approximate solution is piecewise planar, but continuous everywhere.

The coefficients a, b, c in Eq. (3.01) may be found from the three independent simultaneous equations which are obtained by requiring the potential to assume vertex values U_1, U_2, U_3 at the three vertices. Substituting the three vertex potentials and locations into Eq. (3.01) in turn, there is obtained

$$\begin{bmatrix} U_1 \\ U_2 \\ U_3 \end{bmatrix} = \begin{bmatrix} 1 & x_1 & y_1 \\ 1 & x_2 & y_2 \\ 1 & x_3 & y_3 \end{bmatrix} \begin{bmatrix} a \\ b \\ c \end{bmatrix}. \tag{3.02}$$

The determinant of the coefficient matrix in Eq. (3.02) may be recognised on expansion as equal to twice the triangle area. Except in the degenerate

Fig. 1.2. Typical triangular finite element in x–y plane.

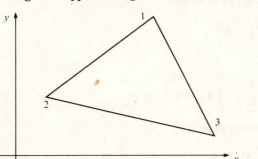

case of zero area, the coefficients a, b, c are therefore readily determined by solving the simultaneous equations, Eqs. (3.02). Substitution of the result into Eq. (3.01) then yields

$$U = \begin{bmatrix} 1 & x & y \end{bmatrix} \begin{bmatrix} 1 & x_1 & y_1 \\ 1 & x_2 & y_2 \\ 1 & x_3 & y_3 \end{bmatrix}^{-1} \begin{bmatrix} U_1 \\ U_2 \\ U_3 \end{bmatrix}. \tag{3.03}$$

Combining x, y and the elements of the inverted coefficient matrix into new functions of position, Eq. (3.03) may be written

$$U = \sum_{i=1}^{3} U_i \alpha_i(x, y), \tag{3.04}$$

where

$$\alpha_1 = \frac{1}{2A}\{(x_2 y_3 - x_3 y_2) + (y_2 - y_3)x + (x_3 - x_2)y\} \tag{3.05}$$

is a linear function of position only, and A represents the surface area of the triangle. The remaining two functions are obtainable by cyclic interchange of subscripts. It is readily verified from (3.05) that the newly defined functions are *interpolatory* on the three vertices of the triangle, i.e., that each function vanishes at all vertices but one, and that it has unity value at that one:

$$\alpha_i(x_j, y_j) = 0 \quad i \neq j$$
$$= 1 \quad i = j. \tag{3.06}$$

The energy associated with a single triangular element may now be determined using Eq. (2.02), the region of integration being the element itself. The potential gradient within the element may be found from Eq. (3.04) as

$$\nabla U = \sum_{i=1}^{3} U_i \nabla \alpha_i, \tag{3.07}$$

so that the element energy becomes

$$W^{(e)} = \tfrac{1}{2} \int |\nabla U|^2 \, dS. \tag{3.08}$$

or, from (3.07),

$$W^{(e)} = \tfrac{1}{2} \sum_{i=1}^{3} \sum_{j=1}^{3} U_i \int \nabla \alpha_i \cdot \nabla \alpha_j \, dS \, U_j \tag{3.09}$$

For brevity, define matrix elements

$$S_{ij}^{(e)} = \int \nabla \alpha_i \cdot \nabla \alpha_j \, dS, \tag{3.10}$$

where the superscript identifies the element. Equation (3.09) may thus be written as the matrix quadratic form

$$W^{(e)} = \tfrac{1}{2}\mathbf{U}^{T}\mathbf{S}^{(e)}\mathbf{U}. \tag{3.11}$$

Here, $\mathbf{U}$ is the column vector of vertex values of potential; the superscript T denotes transposition.

For any given triangle, the matrix $\mathbf{S}$ is readily evaluated. On substitution of the general equation, Eq. (3.05), into Eq. (3.10), a little algebra yields

$$S^{(e)}_{12} = \frac{1}{4A}\{(y_2 - y_3)(y_3 - y_1) + (x_3 - x_2)(x_1 - x_3)\}, \tag{3.12}$$

and similarly for other entries of the matrix $\mathbf{S}$.

4. Element assembly

For any one triangular element, the element energy may be approximately computed as shown above. The total energy associated with an assemblage of many elements is, in general, the sum of all the individual element energies,

$$W = \sum_{\text{all } e} W^{(e)}. \tag{4.01}$$

Any composite model made up of triangular patches may be built up one triangle at a time. It therefore suffices to consider how continuity is enforced when one triangular element is added to an already existing assemblage. For simplicity, suppose the existing assemblage consists of the one triangle 1–2–3 of Fig. 1.3(a), and the triangle 4–5–6 is to be joined to it. Since three potential values are associated with each triangle,

Fig. 1.3. (a) Adjacent triangular elements, considered to be electrically disjoint. (b) Adjacent triangular elements, with potentials required to be continuous, and nodes renumbered accordingly.

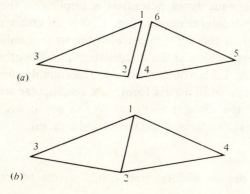

all possible states of the pair of elements are describable by a column vector containing all six vertex potentials,

$$\mathbf{U}_{dis}^{T} = [U_1 \, U_2 \, U_3 \, U_4 \, U_5 \, U_6]_{dis},\tag{4.02}$$

where the subscript 'dis' indicates that disjoint elements (elements as yet not joined together in any way) are being considered. The total energy of the pair of elements is then

$$W = \tfrac{1}{2}\mathbf{U}_{dis}^{T}\mathbf{S}_{dis}\mathbf{U}_{dis},\tag{4.03}$$

where

$$\mathbf{S}_{dis} = \begin{bmatrix} S_{11}^{(1)} & S_{12}^{(1)} & S_{13}^{(1)} & & & \\ S_{21}^{(1)} & S_{22}^{(1)} & S_{23}^{(1)} & & & \\ S_{31}^{(1)} & S_{32}^{(1)} & S_{33}^{(1)} & & & \\ & & & S_{44}^{(2)} & S_{45}^{(2)} & S_{46}^{(2)} \\ & & & S_{54}^{(2)} & S_{55}^{(2)} & S_{56}^{(2)} \\ & & & S_{64}^{(2)} & S_{65}^{(2)} & S_{66}^{(2)} \end{bmatrix}\tag{4.04}$$

is the matrix $\mathbf{S}$ (the *Dirichlet matrix*) of the disjoint pair of elements. More briefly, in partitioned matrix form,

$$\mathbf{S}_{dis} = \begin{bmatrix} \mathbf{S}^{(1)} & 0 \\ 0 & \mathbf{S}^{(2)} \end{bmatrix}.\tag{4.05}$$

In the connected assembly of elements, potential values are physically required to be continuous across interelement boundaries. Because the potential in each triangle is approximated by a linear function of x and y, along any one triangle side, its value varies linearly with distance. Hence the continuity requirement on potentials is satisfied, provided the potentials at corresponding vertices are identical. That is to say, the potential in Fig. 1.3(*b*) will be continuous, if the potentials at vertices 1 and 6 are forced to be exactly equal, and so are the potentials at vertices 2 and 4. Such equality of potentials is implicit in the node numbering for the quadrilateral region, Fig. 1.3(*b*). Of course, there need not be any particular relationship between the node numbers for the triangles, on the one hand, and the quadrilateral, on the other; the numberings shown in Fig. 1.3 are quite arbitrary. The equality constraint at vertices may be expressed in matrix form, as a rectangular matrix $\mathbf{C}$ relating potentials of the disjoint elements to the potentials of the conjoint set of elements (also termed the connected system):

$$\mathbf{U}_{dis} = \mathbf{C}\mathbf{U}_{con},\tag{4.06}$$

where the subscripts denote disjoint and conjoint sets of elements respectively. With the point numberings shown in Fig. 1.3, this equation takes the form

$$
\begin{bmatrix} U_1 \\ U_2 \\ U_3 \\ U_4 \\ U_5 \\ U_6 \end{bmatrix}_{dis} = \begin{bmatrix} 1 & & & \\ & 1 & & \\ & & 1 & \\ & & 1 & \\ & & & 1 \\ 1 & & & \end{bmatrix} \begin{bmatrix} U_1 \\ U_2 \\ U_3 \\ U_4 \end{bmatrix}_{con} . \tag{4.07}
$$

In Eq. (4.07) all those matrix elements have been omitted which must always be zero as a consequence of there being no connection between the corresponding vertices. Substituting Eq. (4.06) into (4.03), the energy for the connected problem becomes

$$ W = \tfrac{1}{2} \mathbf{U}_{con}^T \mathbf{S} \mathbf{U}_{con}, \tag{4.08} $$

where

$$ \mathbf{S} = \mathbf{C}^T \mathbf{S}_{dis} \mathbf{C} \tag{4.09} $$

represents the assembled coefficient matrix of the connected problem. For the assembly in Fig. 1.3,

$$
\mathbf{S} = \begin{bmatrix} S_{11}^{(1)} + S_{66}^{(2)} & S_{12}^{(1)} + S_{64}^{(2)} & S_{13}^{(1)} & S_{65}^{(2)} \\ S_{21}^{(1)} + S_{46}^{(2)} & S_{22}^{(1)} + S_{44}^{(2)} & S_{23}^{(1)} & S_{45}^{(2)} \\ S_{31}^{(1)} & S_{32}^{(1)} & S_{33}^{(1)} & 0 \\ S_{56}^{(2)} & S_{54}^{(2)} & 0 & S_{55}^{(2)} \end{bmatrix} . \tag{4.10}
$$

The disjoint and conjoint numberings are frequently also termed *local* and *global* numberings respectively.

5. Solution of the connected problem

In the above two sections, the energy of a continuous approximate potential distribution was formulated as a quadratic form involving the column vector of node potentials. To obtain an approximate solution of Laplace's equation, it remains to minimise the stored energy in the connected finite element model. Since the energy expression of Eq. (4.08) is quadratic in the nodal potentials, it must have a unique minimum with respect to each component of the potential vector $\mathbf{U}$. Hence, to minimise it is sufficient to set

$$ \frac{\partial W}{\partial U_k} = 0. \tag{5.01} $$

Here the index k refers to entries in the connected potential vector or, what is equivalent, to node numbers in the connected model. Differentiation in Eq. (5.01) with respect to each and every k thus corresponds to an unconstrained minimisation, with the potential allowed to vary at every node. The unconstrained minimisation, however, does not correspond to the boundary-value problem as originally stated in Fig. 1.1. Indeed, the unconstrained minimum energy is trivially zero, with exactly zero potential everywhere. In the boundary-value problem to be solved, certain boundary segments have specified potential values; thus, a certain subset of the potentials contained in the vector **U** must assume exactly those prescribed values. Suppose the node numbering in the connected model is such that all nodes whose potentials are free to vary are numbered first, all nodes with prescribed potential values last. In Fig. 1.1(a), for example, the nodes in the interconductor space (where the potential is to be determined) would be numbered first, and all nodes lying on conductor surfaces (where the potential is prescribed) subsequently. Equation (5.01) may then be written with the matrices in partitioned form,

$$\frac{\partial W}{\partial U_k} = \frac{\partial}{\partial [\mathbf{U}_f]_k} [\mathbf{U}_f^T \mathbf{U}_p^T] \begin{bmatrix} \mathbf{S}_{ff} & \mathbf{S}_{fp} \\ \mathbf{S}_{pf} & \mathbf{S}_{pp} \end{bmatrix} \begin{bmatrix} \mathbf{U}_f \\ \mathbf{U}_p \end{bmatrix} = 0, \tag{5.02}$$

where the subscripts f and p refer to nodes with free and prescribed potentials respectively. Note that, since the prescribed potentials cannot vary, differentiation with respect to them is not possible. Performing the differentiation with respect to the free potentials only, there results the rectangular matrix equation

$$[\mathbf{S}_{ff} \quad \mathbf{S}_{fp}] \begin{bmatrix} \mathbf{U}_f \\ \mathbf{U}_p \end{bmatrix} = 0. \tag{5.03}$$

The rectangular coefficient matrix in this equation contains as many rows as there are unconstrained (free) variables; but its number of columns equals the total number of free as well as prescribed nodal potentials. Rewriting, Eq. (5.03) assumes the form

$$\mathbf{S}_{ff} \mathbf{U}_f = -\mathbf{S}_{ff} \mathbf{U}_p. \tag{5.04}$$

The left-hand coefficient matrix is square and, in general, nonsingular; a formal solution to the problem is therefore given by

$$\mathbf{U} = \begin{bmatrix} -\mathbf{S}_{ff}^{-1} \mathbf{S}_{fp} \mathbf{U}_p \\ \mathbf{U}_p \end{bmatrix}. \tag{5.05}$$

The approximate solution as calculated takes the form of a set of nodal potential values. However, it is important to note that the finite element solution is uniquely and precisely defined everywhere, not only at the triangle vertices, because the energy minimisation assumes the solution surface to have a particular shape. The set of nodal potential values is merely a compact representation for the piecewise-planar solution surface which yields minimum energy.

Within each triangle the local potential values are prescribed by Eq. (3.01). Thus, no further approximation is necessary to obtain contour plots of equipotential values, to calculate the total stored energy, or to perform any other desired further manipulations. Since in this method the potential in each element is taken to be the linear interpolate of its vertex values, as in Eq. (3.04), an equipotential plot will necessarily consist of piecewise-straight contours. For example, Fig. 1.4 shows equipotential contour plots for the problem of Fig. 1.1(a).

It is worth noting in Fig. 1.4 that the Dirichlet boundary conditions (i.e., boundary conditions requiring the potential to take on prescribed values) are exactly satisfied because the potential values at the boundary nodes are explicitly specified when Eqs. (5.03) and (5.04) are set up. On the other hand, the homogeneous Neumann boundary condition (i.e., the requirement of zero normal derivative) is not satisfied exactly, but only in a certain mean-value sense which causes the contour integral term in Eq. (2.04) to vanish. This boundary condition could of course be rigidly imposed; but the stored field energy would be raised thereby, so that the overall solution accuracy would in fact be worse. This point

Fig. 1.4. Equipotential contours for the coaxial-line problem of Fig. 1.1(a).

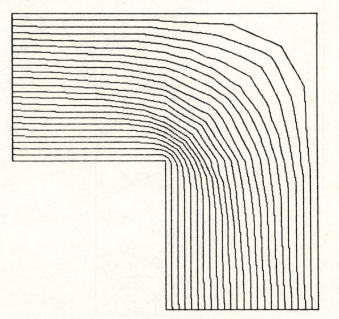

will be considered in some detail later. Roughly speaking, the procedure followed here trades error incurred along the Neumann boundary for an accuracy increase in the solution region.

6.   Poisson's equation

Where distributed sources occur within the field region, an approach similar to the above may be used, but with the difference that the source distributions must be explicitly included. As a simple example, Fig. 1.5 shows an electric machine conductor lying in a slot. It can be readily shown that the magnetic field in the slot can be described by the magnetic vector potential A, which satisfies a vector form of the Poisson equation. If the slot and conductor are assumed infinitely long, both the

Fig. 1.5. Electric machine rotor slot and its triangular finite element representation.

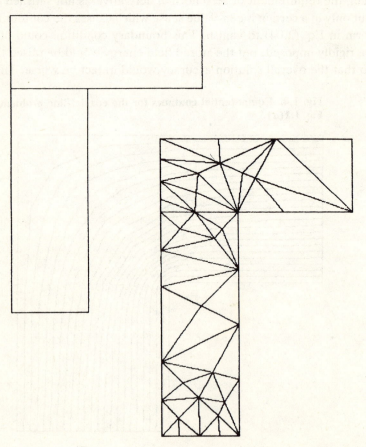

current density $\mathbf{J}$ and the magnetic vector potential $\mathbf{A}$ possess only longitudinally directed components; the vector Poisson equation degenerates to its scalar counterpart:

$$\nabla^2 A = -\mu_0 J. \tag{6.01}$$

If the machine iron is assumed infinitely permeable, the normal derivative of A at the slot centreline and at all iron surfaces must be zero. Further, any constant value of A denotes a flux line. Boundary conditions for the problem are thereby clearly defined. The variational problem equivalent to solving Poisson's equation is that of minimising the energy-related functional

$$F(u) = \tfrac{1}{2} \int |\nabla u|^2 \, \mathrm{d}S - \mu_0 \int uJ \, \mathrm{d}S. \tag{6.02}$$

To show that this functional reaches a minimum at the true solution of Eq. (6.01), suppose A is the correct solution, and v is some differentiable function which vanishes at all boundary points where A is prescribed. Let $F(A + \theta v)$ be evaluated, where θ is a numerical parameter. There is obtained

$$F(A + \theta v) = F(A) + \theta \int \nabla A \cdot \nabla v \, \mathrm{d}S$$
$$- \theta \mu_0 \int vJ \, \mathrm{d}S + \tfrac{1}{2}\theta^2 \int |\nabla v|^2 \, \mathrm{d}S. \tag{6.03}$$

Using Green's theorem once again, the second term on the right may be changed to read

$$\int \nabla A \cdot \nabla v \, \mathrm{d}S = \oint v \frac{\partial A}{\partial n} \, \mathrm{d}s - \int v \nabla^2 A \, \mathrm{d}S. \tag{6.04}$$

The contour integral on the right vanishes since either v or the normal derivative of A is zero at every point along the boundary. Since A is the correct solution of Eq. (6.01), it is further possible to rewrite Eq. (6.04), altering the right-hand surface-integral term, as

$$- \int v \nabla^2 A \, \mathrm{d}S = \mu_0 \int vJ \, \mathrm{d}S. \tag{6.05}$$

The functional of Eq. (6.03) thus simplifies to

$$F(A + \theta v) = F(A) + \tfrac{1}{2}\theta^2 \int |\nabla v|^2 \, \mathrm{d}S. \tag{6.06}$$

Since the integral on the right is always positive, it is evident that a minimum will be reached when θ has zero value; and conversely, that $F(u)$ reaches its minimum value for $u = A$, the solution of Eq. (6.01).

It will be noted that the field energy is still calculable by the general expression

$$W = \frac{1}{2} \int |\nabla A|^2 \, dS \qquad (6.07)$$

or, alternatively, by evaluating the equivalent expression

$$W = \frac{\mu_0}{2} \int AJ \, dS. \qquad (6.08)$$

At its minimum value $F(A)$, F evidently has a negative value equal in magnitude to the total stored energy. Once again, the error term in Eq. (6.06) depends on the square of the parameter θ. Near the correct solution, θ is small. The accuracy with which the stored energy can be found is therefore very high, even if the potential values are locally not very accurate.

7.   Modelling the source term

To construct a finite element model of the Poisson-equation problem, a procedure will be employed similar to that used for Laplace's equation. The problem region will again be triangulated, as shown in Fig. 1.5 and, initially, a typical triangular element will be examined in isolation from the rest. Since the first term in the functional of Eq. (6.02) is identical to the right-hand side of Eq. (2.02), the discretisation process follows exactly the same steps, and leads to exactly the same result for the connected model, Eq. (4.08). The second term in Eq. (6.02) requires treatment which is similar in principle, but slightly different in details.

Over any one triangle, the prescribed current density $J(x, y)$ will be approximated in a manner similar to the potential

$$J(x, y) = \sum_{i=1}^{3} J_i \alpha_i(x, y), \qquad (7.01)$$

where the right-hand coefficients are vertex values of current density within the triangle. These values are of course known, since the current density itself is a prescribed function. The source integral may therefore be written

$$\int AJ \, dS = \sum_{i=1}^{3} \sum_{j=1}^{3} A_i \int \alpha_i \alpha_j \, dS \, J_j, \qquad (7.02)$$

with the vertex potential values the only unknowns. For each element, let another square matrix of order 3 be defined by

$$T_{ij}^{(e)} = \int \alpha_i \alpha_j \, dS, \qquad (7.03)$$

so that

$$\int AJ \, dS = \mathbf{A}^{T} \mathbf{T}^{(e)} \mathbf{J}, \tag{7.04}$$

where the superscript e identifies the element, and the region of integration is understood to be that element.

For the disjoint set of triangular elements, the functional of Eq. (6.02) now becomes

$$F(A) = \tfrac{1}{2} \mathbf{A}_{dis}^{T} \mathbf{S}_{dis} \mathbf{A}_{dis} - \mu_0 \mathbf{A}_{dis}^{T} \mathbf{T}_{dis} \mathbf{J}_{dis}. \tag{7.05}$$

The element interconnection once again expresses itself in the requirement of potential continuity and hence in a constraint transformation like Eq. (4.06). Thus

$$F(A) = \tfrac{1}{2} \mathbf{A}^{T} \mathbf{S} \mathbf{A} - \mu_0 \mathbf{A}^{T} \mathbf{C}^{T} \mathbf{T}_{dis} \mathbf{J}_{dis}. \tag{7.06}$$

Minimisation of $F(A)$ with respect to each and every unconstrained vertex potential, putting

$$\frac{\partial F}{\partial A_k} = 0, \tag{7.07}$$

leads to the matrix equation

$$\mathbf{S}' \mathbf{A} = \mu_0 \mathbf{C}^{T} \mathbf{T}_{dis} \mathbf{J}_{dis} \tag{7.08}$$

as the finite element model of the boundary-value problem. In general, there is no need for source densities to be continuous across interelement boundaries. Therefore, no further transformations need apply to the right-hand side, except in particular cases.

Since differentiation cannot be carried out in Eq. (7.07) with respect to fixed potentials, the matrix $\mathbf{S}'$ of Eq. (7.08) is rectangular. It possesses as many rows as there are unconstrained nodes in the problem, and as many columns as there are nodes in the model. Just as in Eq. (5.03), the vector of node potentials will now be partitioned so as to include all unconstrained potentials in the upper, all prescribed potentials in its lower part. The matrix $\mathbf{S}'$ is partitioned conformably, leading to

$$[\mathbf{S}_{ff} \quad \mathbf{S}_{fp}] \begin{bmatrix} \mathbf{A}_f \\ \mathbf{A}_p \end{bmatrix} = \mu_0 \mathbf{C}^{T} \mathbf{T}_{dis} \mathbf{J}_{dis}. \tag{7.09}$$

As before, the subscripts f and p refer to free and prescribed potential values respectively. Since the latter values are known, they will be moved to the right-hand side:

$$\mathbf{S}_{ff} \mathbf{A}_f = \mu_0 \mathbf{C}^{T} \mathbf{T}_{dis} \mathbf{J}_{dis} - \mathbf{S}_{fp} \mathbf{A}_p. \tag{7.10}$$

Solution of this equation determines the unknown nodal potential values, thus solving the problem.

It is interesting to note that the right-hand side of Eq. (7.10) combines the source term of the differential equation (the inhomogeneous part of the equation) with the effect of prescribed boundary values (i.e., the inhomogeneous part of the boundary conditions). Thus there is no fundamental distinction between the representation of a homogeneous differential equation with inhomogeneous boundary conditions, on the one hand, and an inhomogeneous differential equation with homogeneous boundary conditions, on the other.

A solution, using the rather simple triangulation of Fig. 1.5, appears in Fig. 1.6. Just as in the case of Laplace's equation, the Dirichlet boundary conditions (flux line boundary conditions) are rigidly enforced, while the Neumann boundary conditions are not. As can be seen in Fig. 1.6, the latter conditions are therefore locally violated, but satisfied in the mean.

8.   Practical handling of boundary conditions

The finite element problems as set up in Eqs. (5.03) and (7.09) necessitate numbering the problem variables in a special fashion. To arrange the equations in the manner shown, all fixed potentials must be numbered last, the potentials free to vary must be numbered first. In

Fig. 1.6. Solution of the electric machine slot problem of Fig. 1.5.

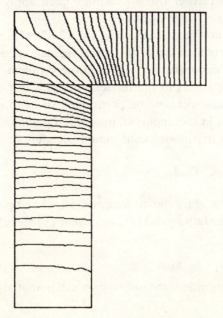

practice, it is not always convenient to renumber variables, nor to partition matrices in this way. Happily, the renumbering and partitioning are only required for purposes of explanation; in practical computing they are never necessary.

Consider again the very simple two-element problem shown in Fig. 1.3(b). It is assumed that potentials 3 and 4 are fixed, 1 and 2 free to vary; in other words, the variable numbering is fully in accordance with the scheme used above. Following (7.09), the matrix equation to be solved then has the form

$$\begin{bmatrix} S_{11} & S_{12} \\ S_{21} & S_{22} \end{bmatrix}\begin{bmatrix} U_1 \\ U_2 \end{bmatrix} = -\begin{bmatrix} S_{13} & S_{14} \\ S_{23} & S_{24} \end{bmatrix}\begin{bmatrix} U_3 \\ U_4 \end{bmatrix}. \tag{8.01}$$

There is not much to be said regarding the high-numbered potentials, which are constrained to have fixed values. Thus the only equation that can be written to describe them is a general form of the identity

$$\mathbf{D}\begin{bmatrix} U_3 \\ U_4 \end{bmatrix} = \begin{bmatrix} D_{33} & \\ & D_{44} \end{bmatrix}\begin{bmatrix} U_3 \\ U_4 \end{bmatrix}, \tag{8.02}$$

where $\mathbf{D}$ is any square, diagonal matrix. Equation (8.02) simply says that the fixed potentials are what they are. It may be combined with Eq. (8.01):

$$\begin{bmatrix} S_{11} & S_{12} & & \\ S_{21} & S_{22} & & \\ & & D_{33} & \\ & & & D_{44} \end{bmatrix}\begin{bmatrix} U_1 \\ U_2 \\ U_3 \\ U_4 \end{bmatrix} = \begin{bmatrix} -S_{13} & -S_{14} \\ -S_{23} & -S_{24} \\ D_{33} & \\ & D_{44} \end{bmatrix}\begin{bmatrix} U_3 \\ U_4 \end{bmatrix}. \tag{8.03}$$

Next, let an arbitrary numbering be introduced for the potentials, one which does not necessarily take the variable and fixed potentials in any particular sequence. For example, let the vertices 1–2–3–4 be renumbered 2–4–1–3, so that the fixed potentials now reside at nodes 1 and 3. The physical problem obviously does not change in any way as a result of the renumbering; only the rows and columns of the coefficient matrix in Eq. (8.03) are permuted into a new sequence in keeping with the new numbering. Thus (8.03) takes the form

$$\begin{bmatrix} D_{11} & & & \\ & S_{22} & & S_{24} \\ & & D_{33} & \\ & S_{42} & & S_{44} \end{bmatrix}\begin{bmatrix} U_1 \\ U_2 \\ U_3 \\ U_4 \end{bmatrix} = \begin{bmatrix} D_{11} & \\ -S_{21} & -S_{23} \\ & D_{33} \\ -S_{41} & -S_{43} \end{bmatrix}\begin{bmatrix} U_1 \\ U_3 \end{bmatrix}. \tag{8.04}$$

The matrix elements left blank in the above are all zero as a consequence of the problem structure, independently of physical dimensions or material properties.

While Eq. (8.04) has more rows and columns than (8.01), so that the cost of solution is higher, there is no need for any particular numbering of vertices and potentials; they may be numbered as desired. Thus the increased cost of handling the matrix problem is at least partly compensated by the work saved in not renumbering and rearranging equations.

In practice, the diagonal matrix $\mathbf{D}$, above, is often taken to be the unit matrix $\mathbf{D} = \mathbf{I}$. Occasionally, matrices $\mathbf{S}$ are encountered whose entries are very large or very small compared to unity, so that numerical roundoff-error considerations may dictate a different choice of $\mathbf{D}$. For example, if the matrix elements in (8.01) are typically of the order of 1.0E−10, they may be lost, compared with unity, on a computer capable of seven-digit arithmetic. Fortunately, such circumstances do not often arise, so that $\mathbf{D} = \mathbf{I}$ is very frequently used. In such a case, (8.04) becomes simply

$$\begin{bmatrix} 1 & & & \\ & S_{22} & & S_{24} \\ & & 1 & \\ & S_{42} & & S_{44} \end{bmatrix}\begin{bmatrix} U_1 \\ U_2 \\ U_3 \\ U_4 \end{bmatrix} = \begin{bmatrix} 1 & \\ -S_{21} & -S_{23} \\ & 1 \\ -S_{41} & -S_{43} \end{bmatrix}\begin{bmatrix} U_1 \\ U_3 \end{bmatrix}. \tag{8.05}$$

Equation (8.05) implies that setting up the finite element equations and imposing the boundary conditions can conveniently be done at the same time, and on an element-by-element basis. As each element matrix is constructed, row and column numbers are scanned to determine whether they correspond to free or fixed potentials. Matrix entries which correspond to free potentials are entered in the natural fashion. Fixed potential values, on the other hand, are treated by substituting rows and columns of the unit matrix on the left, and by augmenting the right-hand side.

9.   Programming and data structures

A major strength of the finite element method, even using first-order triangular elements, resides in its great geometric flexibility. Unlike many other numerical methods, the finite element technique is not strongly restricted in the geometrical shapes which may be treated. Using triangular elements, for example, any two-dimensional region may be treated whose boundary can be satisfactorily approximated by a series of straight-line segments. It should also be noted that the triangular-element mesh by which the interior of the problem region is modelled, is regular neither geometrically nor topologically, the triangles being of varying sizes and shapes, while their interconnection does not necessarily follow a regular pattern.

Approximate solution of a given physical problem by means of finite elements may be regarded as comprising five distinct stages:

(i) creation of finite element mesh, i.e., subdivision of the problem region into elements;

(ii) definition of the sources and imposed boundary values of the problem;

(iii) construction of the matrix representation of each element;

(iv) assembly of all elements, by matrix transformations such as Eq. (4.09), and imposition of boundary conditions;

(v) solution of the resulting simultaneous algebraic equations;

(vi) display and evaluation of the results.

In essence, the geometrically and mathematically complicated boundary-value problem is described as a disjoint set of elements in the first stage; all the subsequent stages serve to reassemble the pieces in a systematic fashion so as to produce the desired solution. The middle three stages clearly involve numerical work of a repetitive and highly systematic character, and are thus ideally suited to the digital computer. The matrix representation of each triangular element can be carried out provided only that the vertex locations of that one triangle are known, without any knowledge of the nature of the entire mesh. Conversely, assembly and imposition of boundary conditions only require knowledge of the mesh topology, i.e., of the manner in which the triangles are interconnected.

Assembly of all individual element matrices to form the global matrix representation requires the connection transformation, Eqs. (4.05)–(4.06), to be executed. All the required topological information is contained in the connection matrix **C**. However, it would clearly be most inefficient to store the connection matrix in the explicit form of Eq. (4.07), and the disjoint global matrices in the form given by Eq. (4.04), for both matrices have rather special form and both contain a high proportion of zero entries. The key to efficient assembly is furnished by Eq. (4.10). The latter may be rewritten in the form

$$
\mathbf{S} = \begin{bmatrix} S_{11}^{(1)} & S_{12}^{(1)} & S_{13}^{(1)} \\ S_{21}^{(1)} & S_{22}^{(1)} & S_{23}^{(1)} \\ S_{31}^{(1)} & S_{32}^{(1)} & S_{33}^{(1)} \end{bmatrix} + \begin{bmatrix} S_{66}^{(2)} & S_{64}^{(2)} & S_{65}^{(2)} \\ S_{46}^{(2)} & S_{44}^{(2)} & S_{45}^{(2)} \\ S_{56}^{(2)} & S_{54}^{(2)} & S_{55}^{(2)} \end{bmatrix}, \quad (9.01)
$$

which suggests that, in the general case, the assembled **S**-matrix may be obtained by calculating each individual element matrix, then immediately adding the nine nonzero matrix contributions due to that element to the corresponding entries in the global **S**-matrix. This technique suggests in turn that a very compact and convenient method of storing the topological information contained in **C** will be by means of an array that identifies the three vertices of each triangle in terms of their global node numbers.

To proceed by way of example, consider the very simple mesh representation of Fig. 1.7 for the slot-conductor problem. The N triangles that constitute the problem region can be identified by a 3-by-N array giving the node numbers, and a 1-by-N array of the source densities in the elements. These arrays may be read in from an input device (card reader, keyboard, etc.). With one element per input line, data for this problem then appear as follows:

1	2	3	1.000
2	4	3	1.000
3	4	5	0.000
4	7	6	0.000
4	6	5	0.000

The assembly then proceeds by zeroing the matrix **S**, forming the element matrix for triangle 1–2–3, adding the nine numbers thus generated to **S**, then computing the element matrix for triangle 2–4–3, adding to **S**, . . . until the entire matrix **S** has been assembled. Computing the matrix representation of each individual element requires knowledge of

Fig. 1.7. A very simple model for the slot problem, to illustrate data handling techniques.

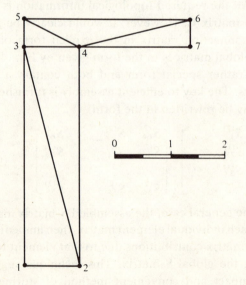

the coordinates of its vertices. For this purpose, 1-by-M arrays of the x- and y-coordinates are needed:

0.000	0.000
1.000	0.000
0.000	4.000
1.000	4.000
0.000	4.500
3.000	4.500
3.000	4.000

Finally, the boundary conditions must be entered. For this purpose, it suffices to specify boundary-point numbers and corresponding boundary values:

6	0.000
7	0.000

The resulting program, implemented in the Fortran language, is given in Section 1.10 below. It will be noted that the program itself contains no information specific to the particular problem being solved. All the geometric, topological, and other problem-dependent information is contained in data arrays, which are prepared separately and read in when required. Another quite different problem in Poisson's equation can be solved using exactly the same program, by supplying a different file of data.

10. A sample program

The methods set out above are embodied in the Fortran program shown below (p. 25). Only the very simplest of methods are employed in this program; but the program structure is essentially identical to many very large, complex programs now in industrial and research use.

The finite element package described here comprises a main program and several subroutines. The main program contains no executable statements other than subroutine calls. It communicates with subroutines through a COMMON block which reappears identically in all the subroutines. In other words, the arrays and other data items are placed in a storage area accessible to all program segments alike. The practical significance of this data arrangement is important: if any alteration is required (for example, if array dimensions are to be changed) it is only necessary to restructure the COMMON block appropriately, and then to replace the COMMON block in every subroutine with an identical copy of the new version. No other reprogramming, and no extensive program testing, will be required.

The main program shown here is purely a steering mechanism: it defines the data structure and sequences the subroutine calls, but it does not actually perform any computations. Such a program organisation is common, for it allows individual subroutines to be replaced without affecting the operation of the others. For example, partial differential equations other than Poisson's equation could be solved by the program shown here – provided the element matrices used were appropriate to the differential equation. But since the element matrices are generated by one subroutine, which does not carry out any other operations, the necessary program alteration consists of removing that one subroutine, and replacing it with a different one. The other program functions, such as data input and equation solving, are unaffected.

The subroutines called by the main program, in the order of their appearance, are:

MESHIN reads in the problem data and stores it in the various data arrays. It does a very limited amount of validity checking. More sophisticated programs of this type often differ in the amount of data verification performed. They are often the largest single subroutines in finite element programs because the number of possible mistakes in a data set is very large. Are there any overlapping triangles? Have all the triangle vertices been properly declared as nodes? Are they all geometrically distinct? Is the problem ill-posed (e.g., all potentials fixed)? Is the problem solvable (e.g., are there any fixed boundary values at all)?

SZERO sets the global element matrices to zero value.

ELMATR computes the matrix representation of one first-order triangular element. It uses techniques discussed in detail in Chapter 3, which differ slightly from those given above.

ELEMB embeds the matrix contributions of one element into the global coefficient matrix and right-hand side.

EQSOLV solves simultaneous algebraic equations, so as to obtain the potentials not known in advance. It assumes that fixed and free potentials may have been intermixed, as discussed in Section 1.8 above, so that it is not necessary to precede EQSOLV by an equation and variable renumbering and rearrangement. EQSOLV is described in detail in Chapter 7.

OUTPUT prints out the solution once it has been obtained. This routine is very complicated in many finite element programs; it often incorporates graphic plotting, calculation of total stored energy, determination of maximal field values, and many other quantities of interest.

In closing, it is worth noting that finite element methods inherently lead to relatively complex programming if reasonable flexibility in problem structure is desired. On the other hand, even at the simplest level, finite element programs can (and should) be written to be highly problem independent. It is important that the user of finite element methods acquire some understanding of the program structure involved, as well as an appreciation of the underlying mathematics. However, it is usually unwise to plunge into *ad hoc* program development for a specific problem – the development of efficient and error-free programs is a complex task often best left to specialists. The user primarily interested in solving particular problems will often find his effort better invested in modifying already existing programs, rather than in succumbing to the temptation to start over again from scratch.

11.   Readings

The finite element method using first-order elements has been applied to a large variety of electrical engineering problems in the past, and will no doubt continue to be applied. Although first-order elements do not produce solutions of high accuracy, the method is simple to understand, simple to program, and above all simple to formulate where the fundamental physical equations are more complicated than those illustrated in this chapter.

The first application of triangular elements to the calculation of electric or other potential fields was probably that of Courant (1943). In his paper, piecewise-approximation methods similar to finite elements were first developed. First-order triangular elements in essentially their present form were developed by Duffin (1959), who indicated the methods for solution and pointed out the availability, not merely of approximate solutions but of bounds for the stored field energy.

There are many monographs and textbooks on finite element theory, as applied to structural engineering. Unfortunately, these are rarely easy for the electrical engineer to read. The text by Norrie & de Vries (1978) is written from the viewpoint of a mechanical engineer, but is sufficiently interdisciplinary to satisfy some electrical engineers as well. The little book by Owen & Hinton (1980) is easy to read, but many will find it not really sufficient for it covers little more than the content of this chapter.

The paucity of textbook material in this field, however, is richly compensated by articles in periodicals and conference records. With the growing popularity of finite element methods for structural analysis in the 1960s, Zienkiewicz & Cheung (1965) attempted solution of practical potential problems and reported on their results. First-order triangular elements were applied to electrical problems shortly thereafter by Silvester (1969), as well as by Ahmed & Daly (1969).

The geometric flexibility of first-order triangles has endeared them to many analysts, so that they are at times used even where higher-order elements would probably yield better accuracy with lower computing costs. Andersen (1973) gives examples of well-working computer programs using first-order elements for daily design work.

In principle, circuits are mathematical abstractions of physically-real fields; nevertheless, electrical engineers at times feel they understand circuit theory more clearly than fields. Carpenter (1975) has given a circuit interpretation of first-order elements.

Particularly if iterative equation-solving methods are employed, first-order finite element techniques sometimes resemble classical finite difference methods. The question is occasionally asked as to which method should then be preferred. This problem was investigated by Demerdash & Nehl (1976), who compared results for sample problems solved by both methods. Finite elements seem preferable even at the first-order level.

An interesting point to note about all finite element methods is that the approximate solution is uniquely defined at all points of interest, not merely on certain discretely chosen points as in finite difference methods. Further use of the solutions, therefore, often requires no further approximation. For example, Daly & Helps (1972) compute capacitances directly from the finite element approximation, without additional assumptions. Similarly, Bird (1973) estimates waveguide losses by reference to the computed solution.

References

Ahmed, S. & Daly, P. (1969). 'Waveguide solutions by the finite element method', *Radio and Electronic Engineer*, **38**, 217–23.

Andersen, O. W. (1973). 'Transformer leakage flux program based on finite element method', *Institute of Electrical and Electronics Engineers Transactions on Power Apparatus and Systems*, **PAS–92**, 682–9.

Bird, T. S. (1973). 'Evaluation of attenuation from lossless triangular finite element solutions for inhomogeneously filled guiding structures', *Electronics Letters*, **9**, 590–2.

Carpenter, C. J. (1975). 'Finite element network models and their application to eddy-current problems', *Institution of Electrical Engineers Proceedings*, **122**, 455–62.

Courant, R. L. (1943). 'Variational method for the solution of problems of equilibrium and vibration', *Bulletin of the American Mathematical Society*, **49**, 1–23.

Daly, P. & Helps, J. D. (1972). 'Direct method of obtaining capacitance from finite element matrices', *Electronics Letters*, **8**, 132–3.

Demerdash, N. A. & Nehl, T. W. (1976). 'Flexibility and economics of the finite element and difference techniques in nonlinear magnetic fields of power devices', *Institute of Electrical and Electronics Engineers Transactions on Magnetics*, **MAG–12**, 1036–8.

Duffin, R. J. (1959). 'Distributed and lumped networks', *Journal of Mathematics and Mechanics*, **8**, 793–826.

Norrie, D. H. & de Vries, G. (1978). *An Introduction to Finite Element Analysis*. New York: Academic Press.

Owen, D. R. J. & Hinton, E. (1980). *A Simple Guide to Finite Elements*. Swansea: Pineridge Press.

Silvester, P. (1969). 'Finite element solution of homogeneous waveguide problems', *Alta Frequenza*, **38**, 313–17.

Zienkiewicz, O. C. & Cheung, Y. K. (1965). 'Finite elements in the solution of field problems', *The Engineer*, Sept. 24, 1965, pp. 507–10.

```
C
C                     A sample Fortran program
C*************** FINITE ELEMENT DEMONSTRATION PROGRAM  ***************
C
C
C
C     THIS PROGRAM READS A PROBLEM FROM FILE "INPUT" AND  SOLVES
C     IT.   THE INPUT IS ECHOED,  AND THE OUTPUT IS WRITTEN,   TO
C     FILE "KONSOL".  THE LOGICAL UNIT NUMBERS ARE ASSIGNED BY A
C     DATA STATEMENT BELOW.
C
C     THE  SUBROUTINES  THAT MAKE UP THIS    PROGRAM  COMMUNICATE
C     VIA A SINGLE COMMON BLOCK.  THE VARIABLES IN COMMON ARE:
C
C        INPUT   =   FORTRAN INPUT DEVICE (OR FILE) NUMBER
C        KONSOL  =   FORTRAN OUTPUT FILE (OR DEVICE) NUMBER
C        IERR    =   ERROR FLAG, ZERO IF ALL IS WELL
C        NVE     =   NUMBER OF NODAL VARIABLES PER ELEMENT
C        MAXNOD  =   MAXIMUM NODE NUMBER (ARRAY DIMENSIONS)
C        MAXELM  =   MAXIMUM NUMBER OF ELEMENTS PERMITTED
C        NODES   =   NUMBER OF NODES USED IN PROBLEM
C        NELMTS  =   NUMBER OF ELEMENTS IN MODEL
C        X, Y    =   NODAL COORDINATES
C        NVTX    =   LIST OF NODES FOR EACH ELEMENT
C        SOURCE  =   SOURCE DENSITY IN EACH ELEMENT
C        CONSTR  =   LOGICAL, .TRUE. FOR FIXED POTENTIALS
C        POTENT  =   NODAL POTENTIAL ARRAY
C        RTHDSD  =   RIGHT-HAND SIDE OF SYSTEM OF EQUATIONS
C        SELM    =   S FOR ONE ELEMENT (WORKING ARRAY)
C        TELM    =   T FOR ONE ELEMENT (WORKING ARRAY)
C        S       =   GLOBAL S-MATRIX FOR WHOLE PROBLEM
C        INTG    =   INTEGER WORKING ARRAY
C
C======================================================================
C
      LOGICAL CONSTR
C
C     THE FOLLOWING ARRAYS HAVE DIMENSIONS EQUAL TO MAXNOD:
      DIMENSION X(50), Y(50), S(50,50)
      DIMENSION CONSTR(50), RTHDSD(50), POTENT(50)
C
C     THE FOLLOWING ARRAYS HAVE DIMENSIONS EQUAL TO MAXELM
C     AND/OR NVE:
      DIMENSION NVTX(3,75), SOURCE(75), SELM(3,3), TELM(3,3),
     1              INTG(3)
```

```
C
      COMMON INPUT, KONSOL, IERR, NVE, MAXNOD, MAXELM, NODES,
     1        NELMTS, X, Y, NVTX, SOURCE, CONSTR, POTENT, RTHDSD,
     2        SELM, TELM, S, INTG
C
C================================================================
C
      DATA MAXNOD, MAXELM, NVE /50, 75, 3/
      INPUT = 5
      KONSOL = 7
C
C     FETCH INPUT DATA FROM INPUT FILE
C
      CALL MESHIN
      IF (IERR .NE. 0) GO TO 900
C
C     SET GLOBAL S-MATRIX AND RIGHT SIDE TO ALL ZEROS.
C
      CALL SZERO
C
C     ASSEMBLE ELEMENT MATRIX.   PROCEED ELEMENT BY ELEMENT.
C
      DO 40 I = 1, NELMTS
C
C          CONSTRUCT ELEMENT S AND T MATRICES
         IE = I
         CALL ELMATR(IE)
C
C          EMBED MATRICES IN GLOBAL S; AUGMENT RIGHT SIDE:
         CALL ELEMBD(IE)
   40    CONTINUE
C
C     SOLVE THE ASSEMBLED FINITE ELEMENT EQUATIONS
C
      CALL EQSOLV
C
C     PRINT OUT THE RESULTING POTENTIAL VALUES
C
      CALL OUTPUT
C
      GO TO 950
  900 WRITE (KONSOL, 1000)
 1000 FORMAT (1X / 1X, 27HCANNOT CONTINUE PROCESSING!)
C
  950 STOP
      END




      1      0.000      0.000
      2      1.000      0.000
      3      0.000      4.000
      4      1.000      4.000
      5      0.000      4.500
      6      3.000      4.500
      7      3.000      4.000

      1  2  3      1.000
      2  4  3      1.000
      3  4  5      0.000
      4  7  6      0.000
      4  6  5      0.000

      6      0.000
      7      0.000
```

```
C
C
C****************************************************************************
C
      SUBROUTINE MESHIN
C
C****************************************************************************
C
C     READS INPUT DATA DECK IN THREE DIVISIONS:  NODES, ELEMENTS,
C     FIXED POTENTIALS.   EACH  DIVISION  IS  FOLLOWED BY  A BLANK
C     LINE, WHICH SERVES AS A TERMINATOR.
C
C     NODES:                    NODE NUMBER, X, Y.   FORMAT I3, 2F10.
C     ELEMENTS:                 NODE NUMBERS, SOURCE.     3I3, F10.
C     POTENTIALS:               NODE NUMBER, FIXED VALUE.   I3, F10.
C
C     ALL DATA ARE ECHOED AS READ BUT LITTLE CHECKING IS DONE FOR
C     VALIDITY.
C
C=================================================================
C     DATA ACCESS COMMON BLOCK -- SAME IN ALL PROGRAM SEGMENTS
C
      LOGICAL CONSTR
      DIMENSION X(50), Y(50), S(50,50)
      DIMENSION CONSTR(50), RTHDSD(50), POTENT(50)
      DIMENSION NVTX(3,75), SOURCE(75), SELM(3,3), TELM(3,3),
     1          INTG(3)
      COMMON INPUT, KONSOL, IERR, NVE, MAXNOD, MAXELM, NODES,
     1       NELMTS, X, Y, NVTX, SOURCE, CONSTR, POTENT, RTHDSD,
     2       SELM, TELM, S, INTG
C=================================================================
C
C     READ IN THE NODE LIST AND ECHO INPUT LINES.
C
C             START BY PRINTING A HEADING.
      WRITE (KONSOL, 1140)
 1140 FORMAT (1X // 8X, 15HINPUT NODE LIST // 3X, 1HN, 8X, 1HX,
     1         11X, 1HY / 1X)
C
C             READ AND ECHO
      NODES = 0
   40 READ (INPUT, 1100) I, XI, YI
 1100 FORMAT (I3, 2F10.5)
      IF (I .NE. 0) WRITE (KONSOL, 1105) I, XI, YI
 1105 FORMAT (1X, I3, 2(2X, F10.5))
C
C             IF NOT END OF NODES (BLANK LINE), INSERT NODE
      IF (I .EQ. 0) GO TO 70
   55 NODES = MAX0 (NODES, I)
      X(I) = XI
      Y(I) = YI
      GO TO 40
C
C             HIGHEST NODE NUMBER ACCEPTABLE?
   60 IF (NODES .LE. MAXNOD) GO TO 70
      WRITE (KONSOL, 1150) NODES, MAXNOD
 1150 FORMAT (1X, 12HNODE NUMBER , I3, 18H EXCEEDS MAXIMUM =, I3)
      GO TO 170
C
C     READ IN THE ELEMENT LIST AND ECHO ALL INPUT AS RECEIVED.
C
C             PRINT HEADING TO START.
   70 WRITE (KONSOL, 1160)
 1160 FORMAT (1X // 6X, 18HINPUT ELEMENT LIST // 3X, 1HI, 5X,
     1         1HJ, 5X, 1HK, 6X, 6HSOURCE / 1X)
```

```
C
C                  READ ELEMENTS IN TURN.   ECHO AND COUNT.
          NELMTS = 0
     80 READ (INPUT, 1170) INTG, SOURCI
   1170 FORMAT (3I3, F10.5)
C
C                  END OF ELEMENT LIST (BLANK LINE) REACHED?
          IF (INTG(1) .EQ. 0) GO TO 120
          WRITE (KONSOL, 1180) INTG, SOURCI
   1180 FORMAT (1X, I3, 2I6, 2X, F10.5)
C
C                  TOTAL NUMBER OF ELEMENTS ACCEPTABLE?
          NELMTS = NELMTS + 1
          IF (NELMTS .LE. MAXELM) GO TO 90
          WRITE (KONSOL, 1190) NELMTS, MAXELM
   1190 FORMAT (15H ELEMENT NUMBER, I4, 18HEXCEEDS MAXIMUM =, I3)
          GO TO 170
C
C                  INSERT NEW ELEMENT IN LIST, GET NEXT ONE.
     90 SOURCE(NELMTS) = SOURCI
          DO 100 M = 1,NVE
    100    NVTX(M,NELMTS) = INTG(M)
          GO TO 80
C
C      READ LIST OF FIXED POTENTIAL VALUES AND PRINT.
C
C                  PRINT HEADER TO START.
    120 WRITE (KONSOL, 1200)
   1200 FORMAT (1X // 4X, 22HINPUT FIXED POTENTIALS // 6X,
        1               4HNODE, 12X, 5HVALUE / 1X)
C
C                  DECLARE ALL NODES TO START OFF UNCONSTRAINED.
          DO 130 M = 1,NODES
             POTENT(M) = 0.
    130      CONSTR(M) = .FALSE.
C
C                  READ AND ECHO INPUT.   CHECK FOR END (BLANK LINE).
    135 READ (INPUT, 1100) I, XI
          IF (I .LE. 0) GO TO 150
          WRITE (KONSOL, 1210) I, XI
   1210 FORMAT (6X, I3, 9X, F10.5)
C
C                  CHECK FOR VALID NUMBER.
          IF (I .LE. NODES) GO TO 140
          WRITE (KONSOL, 1150) I, NODES
          GO TO 170
C
C                  OK -- RECORD VALUE AND GO FOR NEXT ONE.
    140 CONSTR(I) = .TRUE.
          POTENT(I) = XI
          GO TO 135
C
C      RETURN TO CALLING PROGRAM.
C
    150 IERR = 0
          GO TO 900
    170 IERR = 1
    900 RETURN
        END
```

```
C
C****************************************************************
C
      SUBROUTINE SZERO
C
C****************************************************************
C
C     SETS THE GLOBAL S-MATRIX AND THE RIGHT-HAND SIDE RTHDSD TO
C     ALL ZEROS.
C
C================================================================
C     DATA ACCESS COMMON BLOCK -- SAME IN ALL PROGRAM SEGMENTS
C
      LOGICAL CONSTR
      DIMENSION X(50), Y(50), S(50,50)
      DIMENSION CONSTR(50), RTHDSD(50), POTENT(50)
      DIMENSION NVTX(3,75), SOURCE(75), SELM(3,3), TELM(3,3),
     1          INTG(3)
      COMMON INPUT, KONSOL, IERR, NVE, MAXNOD, MAXELM, NODES,
     1          NELMTS, X, Y, NVTX, SOURCE, CONSTR, POTENT, RTHDSD,
     2          SELM, TELM, S, INTG
C================================================================
C
C     SET EVERY ENTRY IN S TO ZERO:
C
      DO 20 I = 1,NODES
         DO 10 J = 1,NODES
            S(I,J) = 0.
   10    CONTINUE
   20 CONTINUE
C
C     SET RIGHT-HAND SIDE VALUES TO ZERO
C
      DO 30 I = 1,NODES
         RTHDSD(I) = 0.
   30 CONTINUE
C
      RETURN
      END




C
C
C****************************************************************
C
      SUBROUTINE ELMATR(IE)
C
C****************************************************************
C
C     CONSTRUCTS THE ELEMENT MATRICES S AND T FOR A SINGLE FIRST-
C     ORDER TRIANGULAR FINITE ELEMENT.   IE = ELEMENT NUMBER.
C
C================================================================
C     DATA ACCESS COMMON BLOCK -- SAME IN ALL PROGRAM SEGMENTS
C
      LOGICAL CONSTR
      DIMENSION X(50), Y(50), S(50,50)
      DIMENSION CONSTR(50), RTHDSD(50), POTENT(50)
      DIMENSION NVTX(3,75), SOURCE(75), SELM(3,3), TELM(3,3),
     1          INTG(3)
      COMMON INPUT, KONSOL, IERR, NVE, MAXNOD, MAXELM, NODES,
     1          NELMTS, X, Y, NVTX, SOURCE, CONSTR, POTENT, RTHDSD,
     2          SELM, TELM, S, INTG
```

```
C==================================================================
C
C       SET UP INDICES FOR TRIANGLE
C
        I = NVTX(1,IE)
        J = NVTX(2,IE)
        K = NVTX(3,IE)
C
C       COMPUTE ELEMENT T-MATRIX
C
        AREA = ABS((X(J) - X(I)) * (Y(K) - Y(I)) -
     1            (X(K) - X(I)) * (Y(J) - Y(I))) / 2.
C
        DO 20 L = 1,NVE
          DO 10 M = 1,NVE
   10       TELM(L,M) = AREA / 12.
   20     TELM(L,L) = 2. * TELM(L,L)
C
C       COMPUTE ELEMENT S-MATRIX
C
        I1 = 1
        I2 = 2
        I3 = 3
C
        DO 30 L = 1,NVE
          DO 30 M = 1,NVE
   30       SELM(L,M) = 0.
C
        DO 50 NVRTEX = 1,3
          CTNG = ((X(J) - X(I)) * (X(K) - X(I)) +
     1         (Y(J) - Y(I)) * (Y(K) - Y(I))) / (2. * AREA)
          CTNG2 = CTNG / 2.
C
          SELM(I2,I2) = SELM(I2,I2) + CTNG2
          SELM(I2,I3) = SELM(I2,I3) - CTNG2
          SELM(I3,I2) = SELM(I3,I2) - CTNG2
          SELM(I3,I3) = SELM(I3,I3) + CTNG2
C
C           PERMUTE ROW AND COLUMN INDICES ONCE
          I4 = I1
          I1 = I2
          I2 = I3
          I3 = I4
          L = I
          I = J
          J = K
          K = L
   50   CONTINUE
C
        RETURN
        END



C
C
C*******************************************************************
C
        SUBROUTINE ELEMBD(IE)
```

```
C
C**********************************************************************
C
C     EMBEDS SINGLE-ELEMENT S AND T MATRICES CURRENTLY IN SELM
C     AND TELM (IN GENERAL COMMON BLOCK) IN THE GLOBAL  MATRIX
C     S.  ARGUMENT IE IS THE ELEMENT NUMBER.
C
C====================================================================
C     DATA ACCESS COMMON BLOCK -- SAME IN ALL PROGRAM SEGMENTS
C
      LOGICAL CONSTR
      DIMENSION X(50), Y(50), S(50,50)
      DIMENSION CONSTR(50), RTHDSD(50), POTENT(50)
      DIMENSION NVTX(3,75), SOURCE(75), SELM(3,3), TELM(3,3),
     1          INTG(3)
      COMMON INPUT, KONSOL, IERR, NVE, MAXNOD, MAXELM, NODES,
     1       NELMTS, X, Y, NVTX, SOURCE, CONSTR, POTENT, RTHDSD,
     2       SELM, TELM, S, INTG
C====================================================================
C
C     RUN THROUGH ELEMENT S AND T MATRICES (SELM AND TELM),
C     AUGMENTING THE GLOBAL S AND THE RIGHT-HAND SIDE AS AP-
C     PROPRIATE.
C
      DO 60 I = 1,NVE
         IROW = NVTX(I,IE)
C
C        DOES ROW CORRESPOND TO A FIXED POTENTIAL?
         IF (CONSTR(IROW)) GO TO 50
C
C        NO, POTENTIAL IS FREE TO VARY.  DO ALL NVE COLUMNS.
         DO 40 J = 1,NVE
            ICOL = NVTX(J,IE)
C
C        DOES COLUMN CORRESPOND TO A FIXED POTENTIAL?
            IF (CONSTR(ICOL)) GO TO 30
C
C        NO; SO AUGMENT S AND RTHDSD.
            S(IROW,ICOL) = S(IROW,ICOL) + SELM(I,J)
            RTHDSD(IROW) = RTHDSD(IROW) + TELM(I,J) * SOURCE(IE)
            GO TO 40
C
C        YES; SO AUGMENT RIGHT SIDE ONLY:
   30       CONTINUE
            RTHDSD(IROW) = RTHDSD(IROW) + TELM(I,J) * SOURCE(IE)
     1                                  - SELM(I,J) * POTENT(ICOL)
   40       CONTINUE
         GO TO 60
C
C        CONSTRAINED ROW NUMBER.  SET GLOBAL S AND RTHDSD.
   50    CONTINUE
         S(IROW,IROW) = 1.
         RTHDSD(IROW) = POTENT(IROW)
C
   60    CONTINUE
C
C     ALL DONE -- RETURN TO CALLING PROGRAM.
      RETURN,
      END
```

```
C
C**********************************************************************
C
      SUBROUTINE OUTPUT
C
C**********************************************************************
C
C     PRINTS THE RESULTS ON THE OUTPUT DEVICE "KONSOL".
C
C====================================================================
C     DATA ACCESS COMMON BLOCK -- SAME IN ALL PROGRAM SEGMENTS
C
      LOGICAL CONSTR
      DIMENSION X(50), Y(50), S(50,50)
      DIMENSION CONSTR(50), RTHDSD(50), POTENT(50)
      DIMENSION NVTX(3,75), SOURCE(75), SELM(3,3), TELM(3,3),
     1          INTG(3)
      COMMON INPUT, KONSOL, IERR, NVE, MAXNOD, MAXELM, NODES,
     1       NELMTS, X, Y, NVTX, SOURCE, CONSTR, POTENT, RTHDSD,
     2       SELM, TELM, S, INTG
C====================================================================
C
C     PRINT THE NODES AND THE OUTPUT POTENTIAL VALUES.
C
      WRITE (KONSOL,1000) (I, X(I), Y(I), POTENT(I),
     1                             I = 1, NODES)
 1000 FORMAT (1X //// 12X, 14HFINAL SOLUTION // 3X, 1HI, 8X, 1HX,
     1         9X, 1HY, 7X, 9HPOTENTIAL // (1X, I3, 2X, F10.5,
     2         F10.5, 3X, F10.5))
C
      RETURN
      END
```

2
Representation of electromagnetic fields

1. Maxwell's equations

Electromagnetic field problems occupy a relatively favourable position in engineering and physics in that their governing laws can be expressed very concisely by a single set of equations, namely those associated with the name of Maxwell. The basic variables are the following set of five vectors and one scalar:

electric field intensity	$\mathbf{E}$	volt/metre
magnetic field intensity	$\mathbf{H}$	ampere/metre
electric flux density	$\mathbf{D}$	coulomb/metre^2
magnetic flux density	$\mathbf{B}$	tesla
electric current density	$\mathbf{J}$	ampere/metre^2
electric charge density	ρ	coulomb/metre^3.

The Maxwell relations may be cast in differential form thus:

$$\nabla \times \mathbf{E} = -\frac{\partial \mathbf{B}}{\partial t}, \tag{1.01}$$

$$\nabla \times \mathbf{H} = \mathbf{J} + \frac{\partial \mathbf{D}}{\partial t}, \tag{1.02}$$

$$\nabla \cdot \mathbf{D} = \rho, \tag{1.03}$$

$$\nabla \cdot \mathbf{B} = 0. \tag{1.04}$$

To these differential relations are added the constitutive relations

$$\mathbf{D} = \varepsilon \mathbf{E}, \tag{1.05}$$

$$\mathbf{B} = \mu \mathbf{H}, \tag{1.06}$$

$$\mathbf{J} = \sigma \mathbf{E}, \tag{1.07}$$

describing the macroscopic properties of the medium being dealt with in terms of permittivity ε, permeability μ and conductivity σ. The quantities ε, μ and σ are not necessarily simple constants, a notable exception being the case of ferromagnetic materials for which the **B**–**H**-relationship may be a highly complicated nonlinear law (cf Chapter 5). Furthermore, ε and μ may represent anisotropic materials, with flux densities differing in direction from corresponding field intensities. In such cases the constitutive constants have to be written as tensors. For free space, using the SI system of units, $\varepsilon_0 = 8.854 \cdots \times 10^{-12}$ farad/metre and $\mu_0 = 4\pi \times 10^{-7}$ henry/metre. The apparently arbitrary value given to μ_0 reflects the basic fact that the units of electromagnetism are adjustable within any framework which preserves the relationship

$$c = (\mu_0 \varepsilon_0)^{-1/2} = 2.998 \cdots \times 10^8 \, \text{m/s},$$

expressing the velocity of light c *in vacuo.* A current may exist under conditions where σ is considered to be zero corresponding to a vacuum but with free electric charge ρ moving with velocity **v**, constituting a current $\mathbf{J} = \rho \mathbf{v}$. In other circumstances materials are dealt with which have infinite conductivity (superconductors) or very high conductivity (copper) in which the current **J** is effectively controlled by external current generators. Equation (1.07) does not describe such situations very usefully. Another fundamental equation expresses the indestructibility of charge:

$$\nabla \cdot \mathbf{J} + \frac{\partial \rho}{\partial t} = 0. \tag{1.08}$$

This *equation of continuity* is consistent with the result of taking the divergence of Eq. (1.02) and substituting it into Eq. (1.03), since the divergence of the curl of any vector vanishes identically. In the truly steady state, of course, Eq. (1.08) becomes $\nabla \cdot \mathbf{J} = 0$. In time-varying situations within good conductors it is also often a good approximation to assume a nondivergent current density. A substantial time-changing flow may be accounted for by virtue of a high density of drifting electrons neutralised by immobile positive ions locked to the lattice of the solid conductor.

1.1   Integral relations

A mathematically equivalent *integral* form of Eqs. (1.01)–(1.04) exists. Integrals are taken over an open surface S or its boundary contour C as illustrated in Fig. 2.1.

The first two of Maxwell's equations become

$$\oint_C \mathbf{E} \cdot d\mathbf{l} = -\int_S \frac{\partial \mathbf{B}}{\partial t} \cdot d\mathbf{S}, \tag{1.09}$$

$$\oint_C \mathbf{H} \cdot \mathbf{dl} = \int_S \left(\mathbf{J} + \frac{\partial \mathbf{D}}{\partial t} \right) \cdot \mathbf{dS}. \tag{1.10}$$

These correspond, respectively, to Faraday's law and Ampere's circuital rule with Maxwell's addition of *displacement current* $\partial \mathbf{D}/\partial t$. The divergence equations, Eqs. (1.03) and (1.04), correspond to Gauss's flux laws

$$\oint_{S'} \mathbf{D} \cdot \mathbf{dS} = \int_\Omega \rho \, \mathrm{d}\Omega, \tag{1.11}$$

$$\oint_{S'} \mathbf{B} \cdot \mathbf{dS} = 0, \tag{1.12}$$

which state that the flux of **B** over any closed surface S' is zero whilst the corresponding integral of **D** equals the total charge within the volume Ω enclosed by the surface S'.

1.2   *Complex phasor notation*

In many circumstances the problem being dealt with concerns the steady state reached when a system is excited sinusoidally. In such cases it is convenient to represent the variables in a complex phasor form. This means that each and any of the electromagnetics variables is represented by a phasor quantity, say in the case of electric field $\mathbf{E}_p = \mathbf{E}_r + \mathrm{j}\mathbf{E}_i$. The corresponding physical, time-varying field is recovered as the real part of $\mathbf{E}_p \exp(\mathrm{j}\omega t)$, where ω is the angular frequency of the sinusoidal excitation. Then when the partial differential operator $\partial/\partial t$ acts upon one of the electromagnetics variables it may be replaced by the simple arithmetic product factor $\mathrm{j}\omega$. Thus Eq. (1.01) becomes

$$\nabla \times \mathbf{E} = -\mathrm{j}\omega \mathbf{B}, \tag{1.13}$$

and so forth. Somewhere within the system being investigated an arbitrary reference of phase will have been established, corresponding to at least one of the spatial components of $\mathbf{E}_p$, E_{p1} say, being purely real. The physical, time-varying quantity here is $E_{p1} \cos(\omega t)$. Elsewhere,

Fig. 2.1. The surface S and contour C for the integral form of Maxwell's equations.

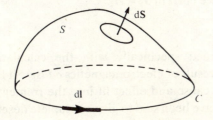

E will generally assume complex values corresponding to a phase shift from the reference of amount ϕ_1, so that

$$E_{p1} = |E_{p1}| \exp{(j\phi_1)} \qquad (1.14)$$

for this particular component. In this case the physical field is $|E_{p1}| \cos{(\omega t + \phi_1)}$. This procedure is an extension of the familiar one-dimensional phasor circuit analysis.

1.3   *Boundary rules*

The laws governing the behaviour of the field vectors at discontinuities in material properties follow from the necessity that the integral forms of the Maxwell equations must be valid for surfaces and contours straddling such boundaries. It is a standard exercise in expositions of electromagnetics theory to show that Eqs. (1.09)–(1.12) are equivalent to a set of rules at material interfaces as follows:

(i) Tangential **E** is always continuous.

(ii) Tangential **H** is discontinuous by an amount corresponding to any surface current **J**$_s$ which may flow.

(iii) Normal **B** is always continuous.

(iv) Normal **D** is discontinuous by an amount corresponding to any surface charge ρ_s which may be present.

The surface current and charge of (ii) and (iv), respectively, are not normally encountered except when one of the materials is a good conductor. At high frequencies there is a well-known effect which confines current largely to surface regions. The so-called skin depth in common conductors is often sufficiently small for the surface phenomenon to be an accurate representation. Thus the familiar rules for the behaviour of time-varying fields at a boundary defined by 'good' conductors follow directly from consideration of the limit when the conductor is 'perfect'. No time-changing field can exist within the conductor, so it must be that:

(i) Electric field is entirely normal to the conductor and is supported by a surface charge $\rho_s = D_n$.

(ii) Magnetic field is entirely tangential to the conductor and is supported by a surface current $J_s = H_t$.

1.4   *Cause and effect*

This text is concerned specifically with the calculation of electromagnetics *effects* as a result of electromagnetics *causes*. It is often not readily appreciated how cause and effect fit into the framework of Maxwell's equations, which have been said to give a complete description of electromagnetics phenomena. Certainly it is easy to see how the specification of a time-varying **J** everywhere would lead to a completely determined field. However, this is but one particularly simple situation. Other electromagnetics 'causes' are often associated with boundary impositions such as the applicaton of voltages to a set of electrodes. Such cases will be discussed in the text individually as they arise.

2.   Potential equations

In many instances a considerable simplification arises both conceptually and in computation if the fields **E** and **B** are represented by *potential functions*. The case of a static electric field being represented by a scalar potential variable

$$\mathbf{E} = -\nabla V,\tag{2.01}$$

where, in charge-free regions, V has to satisfy Laplace's equation

$$\nabla^2 V = 0,\tag{2.02}$$

is probably quite familiar. This is a special case of a more general set of potentials which are now discussed.

The rules of vector calculus require that the curl of the gradient of any scalar variable and the divergence of the curl of any vector variable must each vanish identically. Thus it is readily checked that

$$\mathbf{E} = -\nabla V - \frac{\partial \mathbf{A}}{\partial t},\tag{2.03}$$

$$\mathbf{B} = \nabla \times \mathbf{A},\tag{2.04}$$

define scalar and vector potentials V and **A** respectively, which automatically satisfy two of the four Maxwell relationships, namely Eqs. (1.01) and (1.04). There is evidently an element of arbitrariness in the potential representation of **E** and **B** here, since adding the gradient of *any* scalar function of position ∇C to **A** does not affect the result of Eq. (2.04). A detailed examination of this state of affairs reveals that, because of the arbitrariness described, $\nabla \cdot \mathbf{A}$ may be specified at will. Commonly, the *Lorentz gauge*

$$\nabla \cdot \mathbf{A} = -\mu\varepsilon\frac{\partial V}{\partial t}\tag{2.05}$$

is assumed to hold. Other gauges may be chosen, for instance the Coulomb gauge $\nabla \cdot \mathbf{A} = 0$. However, the Lorentz gauge has the merit that it leads to a simple and symmetrical set of equations when the remaining two of Maxwell's equations, Eqs. (1.02) and (1.03), are cast in terms of the potentials V and $\mathbf{A}$. In the case of a linear, homogeneous medium, these equations become

$$\nabla^2 V - \mu\varepsilon \frac{\partial^2 V}{\partial t^2} = -\rho/\varepsilon, \tag{2.06}$$

$$\nabla^2 \mathbf{A} - \mu\varepsilon \frac{\partial^2 \mathbf{A}}{\partial t^2} = -\mu\mathbf{J}. \tag{2.07}$$

2.1   *Retarded potential solutions*

Equations (2.06) and (2.07) are recognised as wave equations which are generalisations of Poisson's equation with source terms corresponding to charge and current respectively. For localised elements $\rho \, d\Omega$ and $\mathbf{J} \, d\Omega$ it is readily verified that there are the so-called *retarded potential solutions*

$$dV = \frac{[\rho] \, d\Omega}{4\pi\varepsilon r_1}, \tag{2.08}$$

$$d\mathbf{A} = \frac{\mu[\mathbf{J}] \, d\Omega}{4\pi r_1}, \tag{2.09}$$

where r_1 is the magnitude of the radius vector from the point P, at which the potential is required, to the volume element concerned. The square brackets signify that ρ and $\mathbf{J}$ must be specified at a time $r_1(\mu\varepsilon)^{1/2}$ earlier than that for which V or $\mathbf{A}$ is being evaluated, corresponding to the time it takes for an electromagnetic disturbance to proceed from its cause to the point P. It may be noted that in static situations the potentials V and $\mathbf{A}$ are uncoupled, the pairs of equations, Eqs. (2.03) with (2.06) and (2.04) with (2.07), defining $\mathbf{E}$ and $\mathbf{B}$ respectively, now being entirely independent. Equations (2.08) and (2.09) then correspond to the Coulomb and Biot–Savart laws respectively:

$$dE_r = \frac{\rho \, d\Omega}{4\pi\varepsilon r_1^2}, \tag{2.10}$$

$$dH_\phi = \frac{J \, d\Omega \sin\theta}{4\pi r_1^2} \tag{2.11}$$

(see Fig. 2.2 for the geometry relating to Eq. (2.11)). The uncoupled $\mathbf{E}$ and $\mathbf{H}$ remain accurate representations at low frequencies when the displacement current $\partial\mathbf{D}/\partial t$ in Maxwell's equation, Eq. (1.02), and the time difference between source points and places of measurement are negligible.

2.2 Scalar magnetic potential

Another useful potential representation is the magnetic counter-part of Eq. (2.01):

$$\mathbf{H} = -\nabla P, \tag{2.12}$$

where P is a *scalar magnetic potential*. Here the vector calculus rule that the curl of the gradient of any scalar variable is zero gives $\nabla \times \mathbf{H} = 0$. Clearly, because of the Maxwell equation

$$\nabla \times \mathbf{H} = \mathbf{J} + \frac{\partial \mathbf{D}}{\partial t}, \tag{2.13}$$

the representation of magnetic field by a scalar potential is only valid in current-free regions at frequencies sufficiently low for displacement current $\partial \mathbf{D}/\partial t$ to be neglected. Nevertheless, this is often a useful approximation. The basic equation governing scalar magnetic potential derives from the solenoidal property of $\mathbf{B}$, Eq. (1.04), which requires that

$$\nabla \cdot (\mu \nabla P) = 0. \tag{2.14}$$

2.3 Boundary rules for potentials

Throughout this work we will be concerned with the behaviour of fields in subvolumes, 'finite elements', abutting other similar volumes. Sometimes there will be an abrupt change in material properties on passing from one element to another. The rules for the behaviour of the fields themselves at such boundaries have already been noted, being a consequence of Maxwell's laws. It is necessary to establish corresponding rules for the potentials which have just been introduced. Usually V and the components of $\mathbf{A}$ will be taken as continuous across

Fig. 2.2. The geometry relating to the Biot–Savart law, Eq. (2.11).

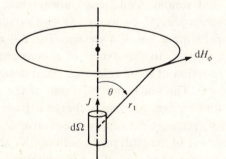

such boundaries. It is readily seen, from the fundamental relations defining fields in terms of potentials, Eqs. (2.03) and (2.04), that this stipulation is *sufficient* to ensure continuity of tangential **E** and normal **B**, as is required under all circumstances by the boundary rules. It is also clear that if there are discontinuities in μ or ε, it will be *necessary* for the normal derivatives of V and **A** to be discontinuous. This is in order that the normal component of $\mathbf{D} = -\varepsilon(\nabla V + \partial \mathbf{A}/\partial t)$ and the tangential component of $\mathbf{H} = (\nabla \times \mathbf{A})/\mu$ should obey their boundary rules (in the absence of surface charge and current, say) of being continuous.

In dealing with 'finite elements' and the extraction of useful field approximations from an assembly of elements, generally the condition of continuity on V and **A** themselves across element boundaries will be imposed as a prerequisite for trial functions before any other constraints are applied. The question then has to be asked as to whether any further measures need actively to be taken with respect to the spatial derivatives of V and **A** when setting up the trial potential functions, in order that the boundary rules should be obeyed. It turns out that constraints are always applied to the trial functions selected so the Maxwell's equations or their variational equivalents, in one form or another, are approximated everywhere in individual elements right up to their boundaries. Since the boundary rules themselves are entirely consequences of Maxwell's equations, it follows that the required behaviour of the boundary-normal derivatives will arise *naturally*. This will occur to the same order of accuracy as is achieved for conformity with respect to Maxwell's equations within the volume concerned.

2.4   *Physical interpretations*

The physical interpretation and visualisation of scalar potential in static cases is familiar and simple. Conductors always take up a constant potential, the difference in V between any two conductors corresponding to the physical application of a voltage. There is always an arbitrariness associated with V, to the extent of a constant. The uncertainty is resolved by adoption of some reference point designated as having zero potential $V = 0$. The scalar potential unit of the *volt* is identifiable as energy per unit charge. Moving a charge q from point X_1 at potential V_1 to X_2 at V_2 requires the supply of energy $q(V_2 - V_1)$ from external sources, regardless of the path taken between X_1 and X_2. Equipotential surfaces of constant V are visualised in conjunction with electric field flux lines to form an orthogonal system. Scalar magnetic potential P is similarly straightforward to appreciate.

The interpretation and visualisation of vector magnetic potential **A** and of V in time-varying situations is not nearly so simple. In a sense this is because potentials are twice removed from reality, once from the electromagnetic fields which are recoverable from them, whilst the fields themselves are merely abstractions representing action at a distance between electric charges and currents. To interpret V under time-varying circumstances we note that work done on a charge q in being moved from X_1 to X_2 is

$$W_{21} = -q \int_{X_1}^{X_2} \mathbf{E} \cdot d\mathbf{l} = q \int_{X_1}^{X_2} \left(\nabla V + \frac{\partial \mathbf{A}}{\partial t} \right) \cdot d\mathbf{l}, \tag{2.15}$$

$$= q(V_2 - V_1) + q \frac{\partial}{\partial t} \int_{X_1}^{X_2} \mathbf{A} \cdot d\mathbf{l}. \tag{2.16}$$

Thus, in general, because of the second term on the right-hand side of Eq. (2.16), W_{21} depends upon the path of integration and the concept of a unique voltage between pairs of points is no longer valid. However, V itself still represents the *conservative* part of the field. A useful property of the vector potential **A** is obtained by applying Stokes's theorem (see for instance Ramo, Whinnery & Van Duzer, 1965, p. 118) to the equation defining magnetic flux through the surface S:

$$\Phi = \int_S (\nabla \times \mathbf{A}) \cdot d\mathbf{S}. \tag{2.17}$$

Equation (2.17) transforms, by means of the theorem, to

$$\Phi = \oint_C \mathbf{A} \cdot d\mathbf{l}, \tag{2.18}$$

where C is the closed curve which bounds any open surface S. Thus the line integral of **A** around a closed circuit gives the magnetic flux threading that circuit. Another characteristic of **A** derives from Eq. (2.09) which shows that the contribution to the vector potential $d\mathbf{A}$ of a current element $\mathbf{J} \, d\Omega$ is in the same direction as **J** itself. Many symmetrical geometries thus result in a vector potential which is aligned exclusively along one of the coordinate axes.

3.   Stationary functionals of potentials

It has already been seen in Chapter 1 how the actual field solution of Laplace's equation in an electrostatics problem corresponds to an electrical stored energy which is smaller than that for all other possible field distributions fitting the Dirichlet boundary conditions of the problem. Possible functions for the potential U were considered and it was asserted from consideration of the minimum stored energy principle that the function of the function U or *functional*,

$$F = \tfrac{1}{2} \int (\nabla U)^2 \, dS, \qquad\qquad (3.01)$$

integrated over the whole two-dimensional problem region had to be stationary about the true solution $U = u$. This *variational* principle was exploited to assemble a patchwork of piecewise-planar functions of the space variables in two dimensions which, within the limitations of the linear approximation, was closest to the correct electrostatics solution. Such a procedure is typical of the methods commonly used for the finite element solution of electromagnetics problems. In general, it may be said that corresponding to any electromagnetic partial differential equation it will be possible to find a functional F. The functional will be expressible in terms of the dependent variables of the equation, whether these are the basic field variables of Section 1 or the potential equivalents of Section 2. The functional F is finally obtained by an integration over the volume concerned in the problem. It will usually have the property of being stationary, either as a maximum, minimum or saddle point, about the correct set of functions representing the required solution to the electromagnetics problem, subject to its boundary constraints. The so-called *calculus of variations* represents a well-established method for solving problems in mathematical physics and engineering. The reader may wish to consult one of the many reference books dealing with the subject in detail, for instance Riley (1974) or Morse & Feshbach (1953). However, at this level we will develop the topic by reference to specific problems in electromagnetism.

3.1   *The inhomogeneous Helmholtz equation*

Following the introductory discourse in Chapter 1 we now consider what may formally be described as the *inhomogeneous Helmholtz equation*

$$\nabla \cdot (p \nabla u) + k^2 u = g, \qquad\qquad (3.02)$$

involving what may be assumed to be a scalar potential variable u. In Eq. (3.02) we take u, a function of position say $u = u(x, y, z)$ referred to a rectangular Cartesian coordinate system, as the dependent variable whose determination constitutes the solution required. The material properties of the medium concerned are supposed represented by $p(x, y, z)$, a specified function of position. The quantity k^2 is a constant, invariant with position, and may or may not be known, whilst $g(x, y, z)$ is a driving function, supposed given. Boundary conditions are specified on a closed surface S, either Dirichlet or homogeneous Neumann, say

u being given on S_1 (Dirichlet), whilst $\partial u/\partial n = 0$ is a specified on S_2 (homogeneous Neumann). Here n represents the space coordinate normal to the surface S, and $\partial u/\partial n$ is the rate of change of u on moving along an outward normal to the surface.

No apologies are made for having posed the dry mathematical problem above. It is part of the finite element 'game' that general equations like (3.02) can be programmed for solution by computer to represent a whole range of different physical situations. One of the important roles to be played by the engineer is in fitting specific problems to such all-embracing packaged programs as are now available. Some examples of various simplifications of Eq. (3.02) occurring in practice follow:

(i) Laplace's equation

$$\nabla^2 u = 0. \tag{3.03}$$

This represents the simple electrostatics problem discussed in Chapter 1 and is implicit in the steady-state, charge-free form of Eq. (2.06). The boundary conditions will often be of the Dirichlet type, the potential u perhaps being specified on electrode boundary surfaces.

(ii) Poisson's equation

$$\nabla \cdot (\varepsilon \nabla u) = -\rho. \tag{3.04}$$

Here, a more complicated potential problem is modelled. The permittivity ε may vary with position in any specified manner. A space-charge distribution ρ, regarded as given, modifies the potential distribution of the simple Laplace situation of (i). The boundary specification would again commonly be Dirichlet.

(iii) Helmholtz equation in two dimensions

$$\nabla_T^2 H_z + k_c^2 H_z = 0. \tag{3.05}$$

The equation here models the behaviour of transverse electric (TE) modes in a waveguide. The formulation of waveguide problems is discussed later in Section 8 of this Chapter. It is sufficient to note at present that ∇_T^2 is the transverse Laplacian operator, $\partial^2/\partial x^2 + \partial^2/\partial y^2$ in rectangular Cartesian systems say, whilst k_c is a cut-off wavenumber, determined by the problem itself through the boundary constraints. The latter in this problem correspond to the homogeneous Neumann condition at the waveguide boundary. Many other examples of physical situations fitting Eq. (3.02) can of course be found.

3.2   *A functional for the Helmholtz equation*

Returning now to Eq. (3.02) we ask whether a functional $F(U)$ can be found which is stationary about the solution $U = u$, the correct solution to this equation and its boundary constraints. The idea would of course be to exploit the stationary property of F in finite element approximations, as was done in the simple examples of Chapter 1. It will be shown that

$$F(U) = \tfrac{1}{2} \int_\Omega [p(\nabla U)^2 - k^2 U^2 + 2gU] \, d\Omega, \tag{3.06}$$

integrated over the whole problem-space, is the required functional, provided that the argument $U(x, y, z)$ is restricted to functions which satisfy the Dirichlet boundary conditions (u specified on S_1, part or whole of the closed boundary surface S). It is to be noticed that no restriction is placed on U with respect to the homogeneous Neumann condition ($\partial u/\partial n = 0$) on S_2, the remaining part of S. Significantly, it turns out that no such restriction is required. This situation is typical in variational analysis. Some boundary conditions, like the Dirichlet restriction here, must be imposed upon trial functions U in order that a particular functional $F(U)$ should become stationary about a true solution u. Such a restriction is described as a *principal boundary condition*. On the other hand, the homogeneous Neumann condition here assumes the role of a *natural boundary condition*. Unconstrained functions U which make $U = u$ a stationary point of $F(U)$ automatically satisfy the condition $\partial u/\partial n = 0$ on S_2.

In order to justify the assertions just made, let

$$U = u + \theta h, \tag{3.07}$$

where θ is some number which can be imagined as varying from a small negative value through zero to a small positive number. Here h is an arbitrary function except that, in order to restrict U to functions satisfying the problem's Dirichlet boundary conditions, h must vanish on S_1. The situation is illustrated for a one-dimensional example in Fig. 2.3.

In order to examine how the functional (3.06) behaves when perturbed about the true solution of (3.02), $u(x, y, z)$, we note that putting $U = u + \theta h$ into Eq. (3.02) yields the following terms:

$$p(\nabla U)^2 = p(\nabla u)^2 + 2\theta p(\nabla u) \cdot (\nabla h) + \theta^2 p(\nabla h)^2, \tag{3.08}$$

$$k^2 U^2 = k^2 u^2 + 2\theta k^2 uh + \theta^2 k^2 h^2, \tag{3.09}$$

$$2gU = 2gu + 2\theta gh. \tag{3.10}$$

Thus, clearly,

$$F(u + \theta h) = F(u) + \theta \int_\Omega [p(\nabla u) \cdot (\nabla h) - k^2 uh + gh] \, d\Omega \tag{3.11}$$

$$+ \tfrac{1}{2}\theta^2 \int_\Omega [p(\nabla h)^2 - k^2 h^2] \, d\Omega.$$

Evidently, the functional chosen as Eq. (3.06) is stationary about $U = u$ if the integral multiplying the first power of θ in Eq. (3.11) vanishes. So we examine the so-called *first variation* in F,

$$\delta F = \theta \int_{\Omega} [p(\nabla u) \cdot (\nabla h) - k^2 uh + gh] \, d\Omega. \tag{3.12}$$

Now, for any vector $\mathbf{X}$, the divergence theorem of vector calculus says that

$$\oint_S \mathbf{X} \cdot d\mathbf{S} = \int_{\Omega} \nabla \cdot \mathbf{X} \, d\Omega, \tag{3.13}$$

so that

$$\oint_S hp(\nabla u) \cdot d\mathbf{S} = \int_{\Omega} \nabla \cdot (hp\nabla u) \, d\Omega$$

$$= \int_{\Omega} [p(\nabla h) \cdot (\nabla u) + h\nabla \cdot (p\nabla u)] \, d\Omega. \tag{3.14}$$

Hence there is an expression

$$\int_{\Omega} p(\nabla h) \cdot (\nabla u) \, d\Omega = \int_S hp(\nabla u) \cdot d\mathbf{S} - \int_{\Omega} h\nabla \cdot (p\nabla u) \, d\Omega, \tag{3.15}$$

which can be used to replace the first term in the integrand of Eq. (3.12), giving

$$\delta F = \theta \int_S hp(\nabla u) \cdot d\mathbf{S} - \theta \int_{\Omega} h[\nabla \cdot (p\nabla u) + k^2 u - g] \, d\Omega. \tag{3.16}$$

Fig. 2.3. (*a*) A trial function $U(x)$ varied through a valid solution $U = u$ subject to a Dirichlet boundary (end) restriction at $x = x_D$ and a homogeneous Neumann condition at $x = x_N$. (*b*) The corresponding arbitrary function $h(x)$, unconstrained at x_N. (*c*) The behaviour of the functional $F(U)$ as θ changes from negative to positive values.

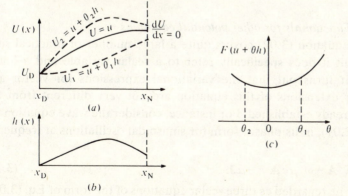

The volume integral in (3.16) above contains the factor $[\nabla \cdot (p\nabla u) + k^2 u - g]$ which, from the definition that u is a solution of Eq. (3.02), vanishes at all points within Ω. Thus, irrespective of the fact that $h(x, y, z)$ is an arbitrary function within Ω, the volume integral of Eq. (3.16) vanishes identically by virtue of its integrand being zero point-by-point. Now, consider the first term of (3.16). It is an integral over the closed surface S comprising S_1 and S_2. The variation of U has been constrained to those functions which fit the principal (Dirichlet) boundary condition, in other words h is zero everywhere on S_1. It can be seen that the factor $(\nabla u) \cdot d\mathbf{S}$ is just $\partial u / \partial n$ at the surface S and this quantity is zero by definition on S_2. Thus, it has been shown that the first variation δF of the functional (3.06) is zero. F is *stationary* about the true solution of the inhomogeneous Helmholtz equation, Eq. (3.02) with its boundary conditions for all arbitrary perturbations, provided the perturbed function satisfies the Dirichlet boundary conditions of the problem, wherever these apply. The homogeneous Neumann boundary restriction has been found to be a natural boundary condition which does not have to be imposed on the trial function.

The procedure always followed when exploiting the variational property just found is to postulate some approximation $u + \theta h$, such as a piecewise-planar function. Such a function itself can never represent the true solution, at least when the piecewise elements are of finite dimensions. However, parameters are chosen within the approximation to ensure that F is stationary. It is evident that the stationary value of $F(u + \theta h)$, now only deviating by a term in θ^2 from the true $F(u)$, is a much more precise estimate of F than is the potential $u + \theta h$ of u. The functional F itself is often closely related to parameters, such as stored energy, which the engineer wishes to determine. Herein lies the power of the variational approach.

3.3   *Functionals for other potential equations*

Equation (3.02) covers quite a large number of practical situations, but it does specifically refer to a scalar variable and a linear system. It turns out that the variational expressions for vector and nonlinear extensions of this equation are not very different from the results already established. For instance, consider the wave equation for $\mathbf{A}$, Eq. (2.07), in its phasor form for sinusoidal oscillations at frequency ω:

$$\nabla^2 \mathbf{A} + \omega^2 \mu \varepsilon \mathbf{A} = -\mu \mathbf{J}. \tag{3.17}$$

This may be regarded as three scalar equations of the form of Eq. (3.02), where u is identified with any one of the vector potential components A_i, say, and g with $-\mu J_i$, the corresponding component of $-\mu \mathbf{J}$. Thus we could expect to be able to apply the variational results for the scalar inhomogeneous Helmholtz equation directly by setting $p = 1$ and $k^2 = \omega^2 \mu \varepsilon$. Some caution, however, is required, since the Laplacian operator acting upon a vector variable conceals a possible trap. If in some coordinate system $A_\xi \hat{\xi}$ is one of the components of $\mathbf{A}$ along a coordinate axis, then the two expressions $\nabla^2(A_\xi \hat{\xi})$ and $(\nabla^2 A_\xi)\hat{\xi}$ are not necessarily the same because in the first only the ∇^2-operator acts upon the unit vector $\hat{\xi}$, which may well vary in direction with the space coordinates. A typical example of this occurs in cylindrical polar coordinates where, at the point (r, ϕ, z) the unit vectors $\hat{\mathbf{r}}$ and $\hat{\boldsymbol{\phi}}$ quite obviously rotate with, and are functions of, r and ϕ. However, the unit vector $\hat{\mathbf{z}}$ in this system is constant so that, nevertheless, $\nabla^2(A_z \hat{\mathbf{z}})$ and $(\nabla^2 A_z)\hat{\mathbf{z}}$ do agree. Of course, the problem does not arise at all in rectangular Cartesian coordinates (x, y, z). Specifically in cylindrical coordinates, the reader may note that the Laplacian operation on a vector must be written as

$$\nabla^2 \mathbf{A} = \left(\nabla^2 A_r - \frac{2}{r^2}\frac{\partial A_\phi}{\partial \phi} - \frac{1}{r^2}A_r\right)\hat{\mathbf{r}}$$

$$+ \left(\nabla^2 A_\phi + \frac{2}{r^2}\frac{\partial A_r}{\partial \phi} - \frac{1}{r^2}A_\rho\right)\hat{\boldsymbol{\phi}} + (\nabla^2 A_z)\hat{\mathbf{z}}, \tag{3.18}$$

where the operator

$$\nabla^2 = \frac{1}{r}\frac{\partial}{\partial r}\left(\frac{r\partial}{\partial r}\right) + \frac{1}{r^2}\frac{\partial^2}{\partial \phi^2} + \frac{\partial^2}{\partial z^2} \tag{3.19}$$

is understood to act straightforwardly only on scalar quantities.

The term $p(\nabla U)^2/2$ in the functional (3.06) may often correspond with a stored energy density. For instance, in the simple electrostatics case of Eq. (3.04), p is associated with permittivity ε and $-\nabla U$ with a trial electric field which becomes the true solution $\mathbf{E}$ in the stationary condition $U = u$. Thus we have

$$\mathcal{E}_\mathrm{E} = \tfrac{1}{2}\varepsilon \mathbf{E}^2 \tag{3.20}$$

for this term, immediately recognisable as stored, electrical energy density in the linear system. If the electrostatics system is nonlinear to the extent of ε being a function of the field $\mathbf{E}$, then standard simple arguments reveal that the stored energy density is now

$$\mathcal{E}_\mathrm{E} = \int_0^\mathbf{E} \varepsilon(\mathbf{e})\mathbf{e} \cdot d\mathbf{e}. \tag{3.21}$$

Usually it may be assumed that if the energy density term in Eq. (3.06) is identified for a particular problem, then its substitution by the non-linear equivalent will yield a correct functional for the nonlinear version of the problem.

4.   Stationary functionals of fields

In some investigations, particularly of high-frequency fields, it is useful to consider functionals written entirely in terms of the fields themselves rather than of potential functions. A typical example would be the determination of the high-frequency fields within a cavity. We would expect to find functionals expressible in terms of the stored electric and magnetic energy densities of the system, $\varepsilon \mathbf{E}^2/2$ and $\mu \mathbf{H}^2/2$ respectively. As a working framework of laws we choose Maxwell's equations in the linear, phasor form which explicitly ignores current $\mathbf{J}$:

$$\nabla \times \mathbf{E} = -j\omega\mu\,\mathbf{H}, \tag{4.01}$$

$$\nabla \times \mathbf{H} = j\omega\varepsilon\,\mathbf{E}. \tag{4.02}$$

However, in this phasor formulation μ and ε may themselves be complex, allowing the possibility, amongst other things, of an implicit current equivalent to Ohm's law

$$\mathbf{J} = \sigma\mathbf{E}. \tag{4.03}$$

It is readily seen how this arises by writing the more complete Maxwell equation applying when the current of Eq. (4.03) is specifically taken into account,

$$\nabla \times \mathbf{H} = \sigma\mathbf{E} + j\omega\varepsilon\,\mathbf{E}. \tag{4.04}$$

Equation (4.04) is observed to be equivalent to

$$\nabla \times \mathbf{H} = j\omega\hat{\varepsilon}\,\mathbf{E}, \tag{4.05}$$

where the complex permittivity

$$\hat{\varepsilon} = \varepsilon - j\sigma/\omega \tag{4.06}$$

arises from taking current flow into account in an otherwise lossless dielectric.

One or other of $\mathbf{E}$ and $\mathbf{H}$ can be eliminated from Eqs. (4.01) and (4.02) to give, alternatively,

$$\nabla \times (\nabla \times \mathbf{H}/\varepsilon) = \omega^2\mu\,\mathbf{H}, \tag{4.07}$$

$$\nabla \times (\nabla \times \mathbf{E}/\mu) = \omega^2\varepsilon\,\mathbf{E}. \tag{4.08}$$

Equations (4.07) and (4.08) demonstrate that any given problem may be worked in terms of *either* $\mathbf{E}$ *or* $\mathbf{H}$ as is preferred. In the form given, the equations allow μ and ε to vary with the space coordinates. Considering the problem first as being worked through **H**, it is asserted that the functional

$$F(\mathbf{H}') = \tfrac{1}{2} \int_{\Omega} [(\nabla \times \mathbf{H}')^2/\omega^2 \varepsilon - \mu \mathbf{H}'^2] \, d\Omega \tag{4.09}$$

is stationary about $\mathbf{H}' = \mathbf{H}$, the true solution of Eq. (4.07), provided the trial function $\mathbf{H}'$ satisfies certain boundary conditions which will shortly be defined. For, let

$$\mathbf{H}' = \mathbf{H} + \mathbf{u}. \tag{4.10}$$

Then the linear term in **u** in the expansion of $F(\mathbf{H} + \mathbf{u})$, the *first variation* of **H**, is given by

$$\delta F = \int_{\Omega} [(\nabla \times \mathbf{H}) \cdot (\nabla \times \mathbf{u})/\omega^2 \varepsilon - \mu \mathbf{H} \cdot \mathbf{u}] \, d\Omega. \tag{4.11}$$

Now exploit the vector identity

$$\nabla \cdot (\mathbf{A} \times \mathbf{B}) = \mathbf{B} \cdot \nabla \times \mathbf{A} - \mathbf{A} \cdot \nabla \times \mathbf{B}, \tag{4.12}$$

so that, identifying $\nabla \times \mathbf{H}/\varepsilon$ with **A** and **u** with **B**, there is obtained

$$-(\nabla \times \mathbf{H}) \cdot (\nabla \times \mathbf{u})/\omega^2 \varepsilon$$
$$= \nabla \cdot [(\nabla \times \mathbf{H}) \times \mathbf{u}/\omega^2 \varepsilon] - \mathbf{u} \cdot \nabla \times (\nabla \times \mathbf{H}/\omega^2 \varepsilon). \tag{4.13}$$

Then Eq. (4.11) becomes

$$\delta F = \int_{\Omega} \{[\nabla \times (\nabla \times \mathbf{H}/\omega^2 \varepsilon) - \mu \mathbf{H}] \cdot \mathbf{u} - \nabla \cdot [(\nabla \times \mathbf{H}) \times \mathbf{u}/\omega^2 \varepsilon]\} \, d\Omega. \tag{4.14}$$

It is observed that the first square-bracketed expression vanishes by virtue of Eq. (4.07) for arbitrary perturbations **u** of the true **H**-field. The divergence theorem can be applied to the remaining term in Eq. (4.14) to give an integral over the boundary surface S,

$$\delta F = - \int_{S} (1/\omega^2 \varepsilon)(\nabla \times \mathbf{H}) \times \mathbf{u} \cdot \mathbf{n} \, dS. \tag{4.15}$$

Hence, from Eq. (4.02),

$$\delta F = \oint_{S} (1/j\omega)(\mathbf{E} \times \mathbf{u}) \cdot \mathbf{n} \, dS. \tag{4.16}$$

From simple vector algebraic rules, Eq. (4.16) may further be written

$$\delta F = \oint_{S} (1/j\omega)(\mathbf{n} \times \mathbf{E}) \cdot \mathbf{u} \, dS, \tag{4.17}$$

$$= \oint_{S} (1/j\omega)(\mathbf{u} \times \mathbf{n}) \cdot \mathbf{E} \, dS. \tag{4.18}$$

Clearly, if the transverse component of **H** is known at the surface *S*, **H**′ = **H** + **u** can be chosen such that **u** × **n** vanishes. A further possibility is that **n** × **E** may vanish, corresponding to the condition holding at a perfect conductor. In this case δ*F* vanishes whatever the perturbation **u** by virtue of Eq. (4.17). It is seen that any piecewise juxtaposition of these two conditions on *S* makes *F* stationary. Practical problems are frequently presented corresponding to such conditions, a transverse **H** being given as the excitation over an aperture in an otherwise perfectly conducting enclosure. Notice that it is not necessary to know the normal field in the aperture, fortunately so because, viewed from the point of view of cause and effect, a transverse field at a cavity aperture is a *cause* which produces the *effect* of internal field and longitudinal aperture field in the cavity. It is important to observe that at perfectly conducting boundaries the functional (4.09), written in terms of the magnetic variable **H**, does not require any constraint to be put on the trial function **H**′ for it to be stationary. We have a *natural*, quasi-Neumann boundary condition here. The specification of **u** × **n** = 0 over a coupling aperture corresponds to a *principal*, quasi-Dirichlet condition. It is noted that, in the functional (4.09), the term $-(\nabla \times \mathbf{H}')^2/\omega^2\varepsilon$ is just $\varepsilon \mathbf{E}'^2$, where **E**′ is an electric field consistent with the trial function **H**′ and Maxwell's equation, Eq. (4.02). It is to be expected that there is a corresponding functional

$$F(\mathbf{E}'') = \tfrac{1}{2} \int_\Omega \left[\varepsilon \mathbf{E}''^2 - (\nabla \times \mathbf{E}'')^2/\omega^2\mu \right] d\Omega, \tag{4.19}$$

which is similarly constructed but expressed entirely in terms of a trial electric field. If **E**″ = **E** + **v** it is clear that now the first variation is reducible to the surface integral

$$\delta F = -\oint_S (1/j\omega)(\mathbf{H} \times \mathbf{v}) \cdot \mathbf{n} \, dS. \tag{4.20}$$

Here, the alternative expressions obtained using the vector algebraic rules are

$$\delta F = -\oint_S (1/j\omega)(\mathbf{n} \times \mathbf{H}) \cdot \mathbf{v} \, dS, \tag{4.21}$$

$$= -\oint_S (1/j\omega)(\mathbf{v} \times \mathbf{n}) \cdot \mathbf{H} \, dS. \tag{4.22}$$

Thus, once again, the contribution to δ*F* vanishes if the transverse component of the field (this time **E**) is supposed known, as at a coupling aperture of a cavity. However, at perfectly conducting boundaries **n** × **H** does not ordinarily vanish, and the trial function **E**″ must be chosen such that it obeys the boundary rules for an electric field there. In practical numerical problems, thus, there is a contrast between using the **H** and **E** alternative dependent variables. On the one hand, **H** at perfectly conducting boundaries is left entirely free whilst, on the other, **E** has to be constrained by a quasi-Dirichlet condition setting the transverse components of **E** to zero.

5. Formulation of potential problems with translational symmetry

Many electromagnetics problems can be approximated by two-dimensional representation corresponding to an infinitely long third axis (say z). Translation in this axial direction reveals no change in material properties, source functions or cross-sectional geometry. Hence, the field solutions themselves are independent of z. There is much to be gained by way of economy in computation when two-dimensional models are used. The general approach to such problems will be illustrated in this section by considering first the electric field in a transmission line of arbitrary cross-section and then the magnetic field in a slot in the iron magnetic circuit of an electric motor armature. In both cases, end effects have to be neglected but, nevertheless, useful engineering parameters can be extracted from such two-dimensional analyses.

5.1 *A coaxial transmission line*

Figure 2.4(a) illustrates one of the simplest possible cases of a two-dimensional potential problem, an inner conductor at potential V_1 completely surrounded by a second conductor at potential V_2. This example has already been preconsidered in Chapter 1 without detailed examination of its formulation from electromagnetics first principles.

Fig. 2.4. (a) A transmission line, illustrating a two-dimensional potential problem. (b) The problem-space reduced by exploiting symmetries and introducing homogeneous Neumann boundary conditions. (c) The transmission line with dielectric-clad inner conductor.

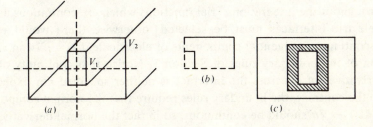

This in fact is very straightforwardly done in terms of a potential function u and the functional

$$F(U) = \tfrac{1}{2} \int_\Omega \varepsilon (\nabla U)^2 \, d\Omega. \tag{5.01}$$

In Section 2 it was shown that according to Maxwell's equations the potential solution, being time invariant and applying to a charge-free dielectric which is at least of piecewise-constant permittivity, must satisfiy Laplace's equation

$$\nabla^2 u = 0. \tag{5.02}$$

The analysis of Section 3 shows that the functional (5.01) is stationary about $U = u$, the true solution of Eq. (5.02), subject to its boundary conditions. The trial function U must be restricted by the constraint that $U = V_1$ and $U = V_2$ on the two boundary surfaces. As has already been remarked, F is suitable for applications in finite element schemes by exploitation of its stationary properties. Strictly, the integration of Eq. (5.01) refers to a volume Ω. This may be considered to be that between the two transverse planes spaced unit distance apart, so that (5.01) may be replaced by a two-dimensional functional

$$F(U) = \tfrac{1}{2} \int \varepsilon (x, y) [\nabla U(x, y)]^2 \, dx \, dy \tag{5.03}$$

in cases of translational symmetry.

Two planes of symmetry exist in the arrangement shown in Fig. 2.4(a). It is readily appreciated that in such a case the problem of Fig. 2.4(b) can be substituted to some advantage, with a homogeneous Neumann boundary condition being applied at the dotted planes of symmetry. The advantage gained is that of quartering the problem-space (hence allowing greater detail in the solution for given computer limitations) and with the inclusion now of a substantial part of the perimeter subject to the more easily managed natural boundary condition.

In Fig. 2.4(c) the addition of dielectric cladding around the inner transmission-line conductor is shown. The functional (5.01) is now evaluated using the appropriate value of permittivity ε everywhere throughout the integration. Trial functions which are continuous across dielectric interfaces must be selected, otherwise there would result discontinuous tangential components of electric field $\mathbf{E} = -\nabla u$ in violation of the boundary rules of Section 1. Notice that continuity of the derivatives of u across the interface is neither specified nor expected. To the contrary, the boundary rules require that the normal component of $\mathbf{D} = -\varepsilon \nabla u$ should be continuous, so in fact the normal derivatives of any true solution u will necessarily be discontinuous across dielectric boundaries. As was explained in Section 3, the interface requirement of normal **D** appears as a *natural* boundary condition.

5.2   *A parallel strip-line system*

Figure 2.5 shows a variant of the transmission-line problem, depicting a parallel-strip system extending to infinity along a second axis as well as into the plane of the diagram. The boundary surface chosen is closed by the dotted planes X_1 and X_2 shown. Provided these are placed well into the region where the conducting electrodes are plane and parallel, the Neumann natural boundary condition again is seen to be appropriate.

5.3   *An open-line pair*

The configuration of Fig. 2.6(a) introduces yet another difficulty. It depicts a pair of strip lines separated by a dielectric sheet but without a closed boundary. Clearly, no process of numerical operations will be able to deal directly with this infinite problem-space. A first approach, often adopted, at overcoming this difficulty is to surround the strip line with a fictitious 'box' at some potential V_3. On the basis that, if far enough away, the box cannot affect the field distribution near the line,

Fig. 2.5. Transmission-line variant, illustrating a further application of the homogeneous Neumann boundary condition.

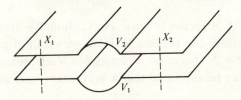

Fig. 2.6. (a) Parallel strips separated by a dielectric sheet. (b) The addition of an arbitrary boundary surface.

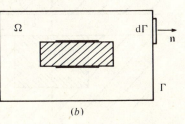

(a)

(b)

the functional F of Eq. (5.01) is evaluated over the truncated problem-space now involved. It is difficult to estimate the error incurred in substituting the finite-space problem except by the clumsy method of reworking the problem for increasingly-distant box boundaries. Some advantage is to be gained by taking V_3 to be a sensible weighted average of V_1 and V_2 (here $(V_1 + V_2)/2$ by symmetry). However, a more sophisticated approach is possible in which the box is treated as a notional boundary outside which there is no interest in detailed knowledge of the field. With reference to Fig. 2.6(b), E signifying the region exterior to the box, it is seen that

$$F(U) = \tfrac{1}{2} \int_{\Omega} \varepsilon (\nabla U)^2 \, d\Omega + \tfrac{1}{2} \int_{E} \varepsilon (\nabla U)^2 \, dE. \qquad (5.04)$$

Attention is focused upon the second volume integral which involves the detailed field knowledge that it has been agreed to dispense with.

A variant of the vector divergence theorem, *Green's theorem*, states that

$$\oint_{S} (\phi \nabla \psi) \cdot d\mathbf{S} = \int_{R} [\phi \nabla^2 \psi + (\nabla \phi) \cdot (\nabla \psi)] \, dR, \qquad (5.05)$$

where S is any surface enclosing a volume R. Putting $\phi = \psi = U$, identifying R with the two-dimensional region E and S with the contour Γ gives

$$\int_{E} [U \nabla^2 U + (\nabla U)^2] \, dE = -\oint_{\Gamma} U(\nabla U) \cdot \mathbf{n} \, d\Gamma$$

$$= -\oint_{\Gamma} U(\partial U/\partial n) \, d\Gamma. \qquad (5.06)$$

The negative sign in Eq. (5.06) arises because, conventionally, the normal $\mathbf{n}$ to S points outwards *from* the volume R. Here we retain $\mathbf{n}$ pointing outwards from Γ, but now this becomes directed *into* R identified as the exterior volume E. It may be supposed that ε is constant in E and that

Fig. 2.7. Armature slot in an electric motor.

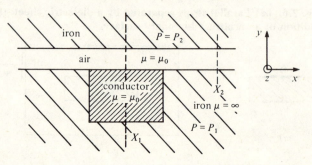

the trial function U has been chosen appropriately such that $\nabla^2 U = 0$ in the exterior region (although it is never necessary to ask precisely what it is). Then

$$F(U) = \tfrac{1}{2} \int_\Omega \varepsilon (\nabla U)^2 \, d\Omega - \oint_\Gamma \varepsilon U \left(\frac{\partial U}{\partial n} \right) d\Gamma. \tag{5.07}$$

Further supposing that ε does not change abruptly at the boundary Γ, $\partial U / \partial n$ may be assumed continuous across Γ. Thus $F(U)$ can be evaluated from Eq. (5.07) by a set of operations which never extends outside the region Ω. By the arguments already advanced, U is stationary about the true solution of Laplace's equation, $U = u$, and so the functional (5.05) can be exploited in finite element schemes.

5.4 *A problem using scalar magnetic potential*

The other example preconsidered in Chapter 1 is that of a slot cut in the armature of a motor (Fig. 2.7) which is supposed to represent detail from the simplified whole motor cross-section depicted in Fig. 2.8. The purpose of the slot is to accommodate a conductor but for the present it is assumed that no current flows and that the magnetic properties of the slot are the same as air, $\mu = \mu_0$. It is shown in elementary accounts of electromagnetics theory that the integral Maxwell–Ampere equation, Eq. (1.10), corresponds, in magnetostatics, to an equation $m = nI$, where m is the *magnetomotive force* (mmf) driving magnetic flux around a magnetic circuit threading an n-turn, current-carrying coil. Moreover, in a similar fashion to *electromagnetic force e* being totalled from changes in scalar electric potential V in electric circuits, m derives from changes in scalar magnetic potential P. A prime consideration in magnetic circuit design is often the minimisation of the *reluctance* of the path. This parameter, analogous to electrical resistance, relates an applied mmf (ampere-turns) to the magnetic flux it produces. To a first approximation the iron of a magnetic circuit can be considered to have infinite permeability. Thus the reluctance associated with the slot may

Fig. 2.8. Magnetic circuit of a simple electric motor.

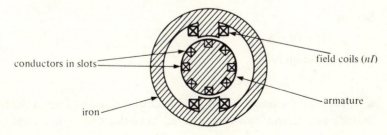

conductors in slots
field coils (nI)
iron
armature

be calculated from its geometry alone, all the mmf due to the nI ampere-turns of the field coil being impressed into the air gap.

Returning to the problem itself we see that it bears considerable resemblance to the electrostatics cases just discussed. Putting $\mu = \mu_0$ for the air and conductor regions, Eq. (2.13) gives $\nabla^2 P = 0$. Thus the functional representing stored magnetic energy,

$$F(Q) = \tfrac{1}{2} \int_\Omega \mu_0 (\nabla Q)^2 \, d\Omega, \tag{5.08}$$

is stationary about the true solution $Q = P$. The boundary conditions are Dirichlet, $P = P_1$ and $P = P_2$, on the iron bounding surfaces, and Neumann, $\partial P / \partial n = 0$, on the plane of symmetry X_1 and at an arbitrary plane X_2 where the field may be considered uniform.

5.5   *A problem using vector magnetic potential*

The formulation used here employing scalar magnetic potential becomes invalid if it is required to take into account current flowing in the slot. The problem has to be reworked in terms of vector potential **A** and now the vector potential problem which arises here when current does flow in the slot-conductor is considered. This current is uni-directional into the plane of the diagram Fig. 2.7, say

$$\mathbf{J} = (0, 0, J(x, y)). \tag{5.09}$$

Since the geometry is translationally symmetric, **A** is to be expected to have the same vector form as **J**, say

$$\mathbf{A} = (0, 0, A(x, y)). \tag{5.10}$$

With μ being piecewise continuous, simplifying Eq. (3.17) and putting $\omega = 0$, gives a valid scalar equation

$$\nabla^2 A = -\mu J, \tag{5.11}$$

which can immediately be applied to the problem. However, it is instructive to derive this equation a little more generally and directly by considering the magnetostatics, Maxwell (Ampere) equation

$$\nabla \times \mathbf{H} = \mathbf{J}. \tag{5.12}$$

Setting

$$\mathbf{H} = (\nabla \times \mathbf{A}) / \mu \tag{5.13}$$

gives immediately

$$\nabla \times [(1/\mu)\nabla \times \mathbf{A}] = \mathbf{J}, \tag{5.14}$$

which is valid for nonuniform μ. If attention is focused upon the translationally-symmetric, two-dimensional but otherwise general case, it can be noted that, after a little vector algebra and defining a *reluctivity* $\nu = 1/\mu$ it follows that

$$\nabla \times \mathbf{A} = \left(\frac{\partial A}{\partial y}, -\frac{\partial A}{\partial x}, 0 \right) \tag{5.15}$$

and

$$\nabla \times (\nu \nabla \times \mathbf{A}) = \left[-\frac{\partial}{\partial x} \left(\nu \frac{\partial A}{\partial x} \right) - \frac{\partial}{\partial y} \left(\nu \frac{\partial A}{\partial y} \right), 0, 0 \right]. \tag{5.16}$$

Thus we arrive at the scalar two-dimensional equation

$$\nabla \cdot (\nu \nabla A) = -J, \tag{5.17}$$

which fits the inhomogeneous Helmholtz equation, Eq. (3.01), precisely, identifying ν with p, $-J$ with g and putting $k^2 = 0$. Thus for linear media, the functional which is stationary about the true solution $U = A$ is, from Eq. (3.06),

$$F(U) = \int_{\Omega} [\nu(\nabla U)^2/2 - JU] \, d\Omega, \tag{5.18}$$

subject to the boundary restrictions as before.

5.6 *An armature slot with a current-carrying conductor*

Returning to the simple system of Fig. 2.7, a second problem, complementary to that of the current-zero system magnetised by field coils external to the immediately-considered volume, can be identified. This second situation concerns nonzero slot-current with zero external magnetisation. Clearly, the variational problem is defined by Eq. (5.18) and all that remains is to identify the boundary constraints applying to this particular geometry. The surface X_1 is a plane of symmetry over which, quite clearly, the magnetic flux vector is to be normal. If the iron is to be considered as having infinite permeability, $\mu = \infty$, flux lines must also emerge normally from it. Otherwise there would have to be a finite tangential **H** within the iron in order to meet the boundary requirement of continuity in this component. This would lead to the impossible situation of there being an infinite magnetic flux density $\mathbf{B} = \mu \mathbf{H}$ within the iron. Observing that

$$\mathbf{B} = \left(\frac{\partial A}{\partial y}, -\frac{\partial A}{\partial x} \right), \tag{5.19}$$

it is immediately seen that the flux normal to any plane implies that $\partial A/\partial n$ is to vanish at that plane. In the two-dimensional form of the problem used here it can be demonstrated that A is constant along any magnetic flux line. For, the change in A on moving from the point (x, y) to $(x + dx, y + dy)$ is

$$dA = \left(\frac{\partial A}{\partial x} dx + \frac{\partial A}{\partial y} dy \right). \tag{5.20}$$

If the displacement (dx, dy) is along a flux line, dx and dy must, respectively, be proportional to the components of $\mathbf{B}$ at (x, y), $\partial A/\partial y$ and $-\partial A/\partial x$. Clearly, dA in Eq. (5.20) vanishes and A has remained constant. Thus, if a plane such as X_2 (Fig. 2.7) can be found which to a good approximation has a uniform magnetic flux density, then it may be assigned an arbitrary constant value, say $A = 0$, to complete the boundary specification for the problem here.

Having solved the two problems concerning the motor slot it would then be possible to combine them. The result would be a useful approximation from which the total magnetic flux density in the slot due to both field-coil excitation and armature current could be estimated.

6.   Formulation of potential problems with axial symmetry

A second class of problems, those possessing axial symmetry, allows a two-dimensional representation with its associated economy in computing resources. Again we consider, as an example, a simple electrostatics case, the field near discontinuity in a coaxial cable, illustrated in Fig. 2.9. A step in the radius of the inner conductor and a dielectric bead are shown as constituting the discontinuity. The functional remains as before, given by Eq. (5.01). The two-dimensional representation now arises because, with azimuthal symmetry, none of the parameters concerned in the analysis depend upon ϕ of the cylindrical coordinate system (r, ϕ, z). Thus ϕ can be immediately integrated out in the volume summation of the functional, the basic element $d\Omega$ becoming a torus of volume,

$$d\Omega = 2\pi r \, dr \, dz, \tag{6.01}$$

Fig. 2.9. Discontinuity on a coaxial cable.

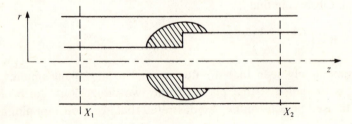

and the functional can be written

$$F(U) = \tfrac{1}{2} \int_{\Omega} \varepsilon(r, z)[\nabla U(r, z)]^2 2\pi r \, dr \, dz. \tag{6.02}$$

Once again, the stationary property of $F(U)$ at the true solution to Laplace's equation, $U = u$, is exploited in finite element schemes. As before, any Dirichlet boundary conditions required need to be imposed beforehand on the trial functions U. Here, these conditions are $U = V_1$ and $U = V_2$ at the inner and outer conductors of the coaxial cable respectively. The problem will no doubt concern a length of cable extending in both directions well away from the discontinuity. The potential gradient is only distorted from the radial direction close to the discontinuity. Thus the problem geometry can be closed by transverse planes X_1 and X_2 shown dotted in Fig. 2.10. The Neumann natural boundary condition may be considered to apply to the planes to a close approximation. As previously, the continuity of tangential electric field at dielectric interfaces requires U to be continuous at the surface of the dielectric bead. The evaluation of functional (6.02) with the appropriate abrupt changes in $\varepsilon(r, z)$ then automatically ensures that the normal field components derived from the stationary property of F satisfy the interface rules of Section 1.

6.1 *An axisymmetric system with nonlinear magnetic material*

Now a corresponding vector potential, axisymmetric problem is examined in which the magnetic field results from currents in solenoidal coils adjacent to an axially-symmetric iron magnetic circuit. A typical practical problem of this type is that of the magnetic lens of an electron microscope (Fig. 2.10). The design of such systems for optimum perform- ance requires very accurate prediction of the magnetic fields for any

Fig. 2.10. Electron microscope magnetic lens.

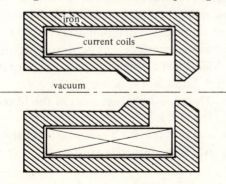

intended lens geometry. As was seen in the translationally-symmetric case, the requirement of including regions where current is flowing dictates that the problem must be worked in terms of vector potential **A**. Rather than attempting to apply the rules which were developed for the essentially scalar Helmholtz equation it is simpler to attack the problem afresh. The treatment of iron magnetic materials as being linear under typical lens-operating conditions is unrealistic. The consideration of nonlinear magnetic problems is dealt with in detail in Chapter 5. However, whilst examining the variational basis for setting up the simpler linear problem nonlinear effects can be included with very little extra complication.

The problem is formulated in terms of a vector potential function **A** so that everywhere the magnetic field is determined by Eq. (2.04), $\mathbf{B} = \nabla \times \mathbf{A}$. It is supposed that the magnetic material of the problem is characterised by an experimentally determined relationship

$$\mathbf{H} = \mathbf{H}(\mathbf{B}). \tag{6.03}$$

Normally the material will be isotropic, that is to say, the direction of **H** is always the same as that of **B**. Also, denoting the magnitude of these two vectors by H and B respectively, it would ordinarily be expected that H would increase monotonically with B. Equation (6.03) may be written in alternative form as

$$\mathbf{H} = \nu(B)\mathbf{B}, \tag{6.04}$$

where ν, the reluctivity (reciprocal of permeability μ) defined in Section 1 is a monotonically increasing scalar function of B. The quantity

$$\mathscr{E}_{\mathrm{M}} = \int_0^{\mathbf{B}} \mathbf{H}(\mathbf{b}) \cdot d\mathbf{b} \tag{6.05}$$

is defined, where a dummy variable **b** has been introduced and Eq. (6.05) signifies that the integral is taken between field limits $\mathbf{b} = 0$ and $\mathbf{b} = \mathbf{B}$. In the simple isotropic case the magnetic material characteristic and the integral (6.05) may be represented diagramatically by Fig. 2.11. The quantity $\mathscr{E}_{\mathrm{M}}$ is recognised as the stored energy density of magnetisation in the iron, although this identification forms no part of the argument to be used here. The problem formulation is completed by assuming that a steady current **J** flows. The time-invariant form of the Maxwell–Ampere equation, Eq. (5.12), $\nabla \times \mathbf{H} = \mathbf{J}$, holds and there will be a unique solution **A** to the boundary-value problem for a given current excitation **J** with constraints placed on **A** at some boundary surface S enclosing a volume Ω. With

$$\mathbf{B}' = \nabla \times \mathbf{A}', \tag{6.06}$$

it is sought to prove that the functional

$$F(\mathbf{A}') = \int_{\Omega} [\mathscr{E}_{M}(\mathbf{B}') - \mathbf{J} \cdot \mathbf{A}'] \, d\Omega \tag{6.07}$$

is stationary about $\mathbf{A}' = \mathbf{A}$. Suppose

$$\mathbf{A}' = \mathbf{A} + \delta \mathbf{A}. \tag{6.08}$$

If

$$\mathbf{B}' = \mathbf{B} + \delta \mathbf{B}, \tag{6.09}$$

where $\mathbf{B}$ is the true solution corresponding to $\mathbf{A}$, then from Eq. (6.06),

$$\delta \mathbf{B} = \nabla \times \delta \mathbf{A}. \tag{6.10}$$

Now

$$\mathscr{E}_{M}(\mathbf{B}') = \int_{0}^{\mathbf{B} + \delta \mathbf{B}} \mathbf{H}(\mathbf{b}) \cdot d\mathbf{b}, \tag{6.11}$$

$$= \mathscr{E}_{M}(\mathbf{B}) + \int_{\mathbf{B}}^{\mathbf{B} + \delta \mathbf{B}} \mathbf{H} \cdot d\mathbf{b}. \tag{6.12}$$

Assume that $\delta \mathbf{B}$ is just a small perturbation on $\mathbf{B}$. The condition of $F(\mathbf{A}')$ being stationary about $\mathbf{A}' = \mathbf{A}$ of course merely requires that the first-order perturbation δF of $F(\mathbf{A})$ should vanish. We observe that to this order in $\delta \mathbf{B}$, Eq. (6.12) may, by elementary rules of calculus, be written

$$\mathscr{E}_{M}(\mathbf{B}') = \mathscr{E}_{M}(\mathbf{B}) + \mathbf{H}(\mathbf{B}) \cdot \delta \mathbf{B}. \tag{6.13}$$

Thus, from Eq. (6.07) there results

$$\delta F = \int_{\Omega} [\mathbf{H} \cdot \delta \mathbf{B} - \mathbf{J} \cdot \delta \mathbf{A}] \, d\Omega. \tag{6.14}$$

Fig. 2.11. *B–H*-curve for a magnetic material.

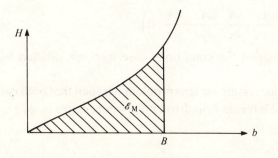

If **J** in Eq. (6.14) is replaced by $\nabla \times \mathbf{H}$, then

$$\delta F = \int_{\Omega} [\mathbf{H} \cdot \delta \mathbf{B} - \nabla \times \mathbf{H} \cdot \delta \mathbf{A}] \, d\Omega \qquad (6.15)$$

and we note that, in arriving at Eq. (6.15), each of the relevant Maxwell laws, Eqs. (1.01), (1.02) and (1.04) has been embodied into the result. Now apply the vector identity

$$\nabla \cdot (\delta \mathbf{A} \times \mathbf{H}) = \mathbf{H} \cdot (\nabla \times \delta \mathbf{A}) - (\nabla \times \mathbf{H}) \cdot \delta \mathbf{A}, \qquad (6.16)$$

from which, using Eq. (6.10), there follows

$$\nabla \cdot (\delta \mathbf{A} \times \mathbf{H}) = \mathbf{H} \cdot \delta \mathbf{B} - \nabla \times \mathbf{H} \cdot \delta \mathbf{A}. \qquad (6.17)$$

Thus Eq. (6.15) becomes

$$\delta F = \int_{\Omega} \nabla \cdot (\delta \mathbf{A} \times \mathbf{H}) \, d\Omega. \qquad (6.18)$$

Equation (6.18) can be transformed using the divergence theorem to give

$$\delta F = \oint_{S} (\delta \mathbf{A} \times \mathbf{H}) \cdot \mathbf{n} \, dS, \qquad (6.19)$$

where **n** is the unit normal to the surface S enclosing Ω. Using the cyclic properties of the vector triple product, Eq. (6.19) can further be written

$$\delta F = \oint_{S} (\mathbf{H} \times \mathbf{n}) \cdot \delta \mathbf{A} \, dS. \qquad (6.20)$$

The first variation of F resulting from a perturbation $\delta \mathbf{A}$ about a true solution **A** evidently vanishes provided one of two conditions holds at the boundary surface S. Either $\mathbf{A}' = \mathbf{A} + \delta \mathbf{A}$ is restricted to functions which satisfy the boundary conditions on S, so that $\delta \mathbf{A}$ itself vanishes there, or S happens to be a surface with its normal coinciding with the direction of **H** so that $\mathbf{H} \times \mathbf{n}$ vanishes. Of course, S may be a composite of surfaces at which one or other of the above restrictions hold. A little algebra reveals that for isotropic media, the $\mathbf{H} \times \mathbf{n} = 0$ condition corresponds to the vector **A** lying in S, also with $\partial \mathbf{A}/\partial n = 0$. For, suppose for the sake of argument, **n** is locally the vector $(0, 0, 1)$. It is found that

$$\mathbf{H} \times \mathbf{n} = \nu \left(\frac{\partial A_x}{\partial z} - \frac{\partial A_z}{\partial x}, \frac{\partial A_z}{\partial y} - \frac{\partial A_y}{\partial z}, 0 \right), \qquad (6.21)$$

which vanishes provided the conditions described are fulfilled locally also.

Returning to the axisymmetric lens problem it is seen that both current and vector potential have, in cylindrical polar coordinates (r, ϕ, z), only ϕ-components

$$\mathbf{J} = (0, J_{\phi}, 0), \qquad (6.22)$$

$$\mathbf{A} = (0, A_\phi, 0). \tag{6.23}$$

On the other hand, the resulting field $\mathbf{B} = \nabla\mathbf{A}$ is directed entirely in the (r, z) plane:

$$\mathbf{B} = \left(-\frac{\partial A_\phi}{\partial z}, 0, \frac{1}{r}\frac{\partial}{\partial r}(rA_\phi) \right). \tag{6.24}$$

The problem is essentially two-dimensional and quasi-scalar in one dependent variable $A(r, z)$, with a driving function $J(r, z)$. The subscript ϕ denoting that A and J are actually components of vectors is suppressed. Clearly, the problem formulation for the linear case is to set up functionals

$$F_E = \int \left[\tfrac{1}{2}\nu\left\{ \left(\frac{\partial A}{\partial z}\right)^2 + \frac{1}{r^2}\left(\frac{\partial}{\partial r}[rA]\right)^2 \right\} - JA \right] 2\pi r\, dr\, dz \tag{6.25}$$

for suitable trial functions A in each finite element Ω_E. The sum of all the individual contributions

$$F = \sum_E F_E \tag{6.26}$$

then can be exploited by virtue of its stationary property when A is the true solution. On the question of boundary conditions and continuity across elements the situation is as in the translationally-symmetric case. Normal $\mathbf{B}$ has to be continuous across each element interface. Since this component is obtained from derivatives of A in the tangential directions, it is sufficient for A to be continuous across such interfaces. There is no requirement that $\partial A/\partial n$ should be continuous. Indeed, continuity of tangential $\mathbf{H} = \nu\mathbf{B}$ requires specific discontinuity of $\partial A/\partial n$ in cases where ν changes abruptly across a boundary. Because the boundary requirements are essentially consequences of Maxwell's laws, the $\partial A/\partial n$ constraints are automatically satisfied when A fits these laws within the bulk of the element.

7.   Wave propagation in uniform guides

The calculation of electromagnetic-wave fields and parameters is frequently required in engineering design and physical analysis. Most electromagnetic-wave-analysis problems can be reduced to the solution of second-order partial differential equations subject to boundary conditions. This fact points to the probable existence of useful techniques for wave calculations being available in the finite element method. In this section the variational formulation of one of the simpler wave-propagation problems, that of the uniform waveguide, is developed in preparation for its finite element treatment in Chapter 3.

This text is primarily concerned with calculations associated with electromagnetics devices rather than with the devices themselves. Thus we might be permitted to set down a few of the basic properties of waveguides and refer the reader to standard works such as Ramo, Whinnery & Van Duzer (1965) or Schelkunoff (1943) for details. Essentially, the behaviour of waveguides derives from the self-evident property of a hollow tube with a highly-reflecting inside surface—that it will channel electromagnetic-wave energy from one end to the other. The immediate example which comes to mind is that of shining light through a tube with a mirror interior surface. Because of the possibility of launching two alternative, independent plane waves, distinct by virtue of possessing orthogonal polarisation, we can expect to find two independent sets of waves or *modes* in the guide. Figure 2.12 shows schematically how a *transverse magnetic* (TM) mode could arise by tilting an axially-directed plane wave in the plane of its **E**-vector. The **H**-vector remains transverse to the axis of the guiding tube. Alternatively, the diagram could be changed to represent the independent, orthogonally polarised wave, the tilting now involving the **H**-vector. Now the **E**-vector remains in place, so that there is a *transverse electric* (TE) wave. However, the waveguide cross-sectional dimensions will invariably be of the same order as the radiation wavelength itself. This considerably complicates the simple picture given. Direct and wall-reflected waves interfere with one another so that within each mode a whole family of waves, corresponding to different orders of interference, will exist. Ultimately, if the dimensions of the guide are too small compared with a wavelength, it will not channel radiation effectively at all.

7.1   *Basic differential equations for waveguide modes*

We assume a cylindrical waveguide of arbitrary cross-section and with its axis aligned along the z-direction. If a coherent wave is to

Fig. 2.12. Schematic synthesis of a TM waveguide mode. The 'wavy' arrowed lines represent two alternative directions of propagation.

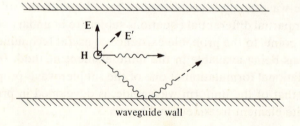

waveguide wall

propagate at frequency ω then the phasor field quantities must take the form

$$E = E(x, y) \exp j(\omega t - \beta z), \tag{7.01}$$

$$H = H(x, y) \exp j(\omega t - \beta z), \tag{7.02}$$

where β is a *propagation constant* corresponding to a wavelength of $2\pi/\beta$ within the guide. More generally, in situations where there is attenuation of the wave the propagation constant is complex and it is customary to write $\gamma = \alpha + j\beta$. Equations (7.01) and (7.02) may be substituted into Maxwell's equations in their current-free form,

$$\nabla \times E = -j\omega\mu H, \tag{7.03}$$

$$\nabla \times H = j\omega\varepsilon E, \tag{7.04}$$

to give

$$\frac{\partial E_z}{\partial y} + j\beta E_y = -j\omega\mu H_x, \tag{7.05}$$

$$-j\beta E_x - \frac{\partial E_z}{\partial x} = -j\omega\mu H_y, \tag{7.06}$$

$$\frac{\partial E_y}{\partial x} - \frac{\partial E_x}{\partial y} = -j\omega\mu H_z, \tag{7.07}$$

$$\frac{\partial H_z}{\partial y} + j\beta H_y = j\omega\varepsilon E_x, \tag{7.08}$$

$$-j\beta H_x - \frac{\partial H_z}{\partial x} = j\omega\varepsilon E_y, \tag{7.09}$$

$$\frac{\partial H_y}{\partial x} - \frac{\partial H_x}{\partial y} = j\omega\varepsilon E_z. \tag{7.10}$$

Equations (7.05) and (7.09) may be solved for H_x and E_y and likewise Eqs. (7.06) and (7.08) for H_y and E_x giving

$$H_x = j\left(\omega\varepsilon \frac{\partial E_z}{\partial y} - \beta \frac{\partial H_z}{\partial x}\right) \bigg/ (k^2 - \beta^2), \tag{7.11}$$

$$H_y = -j\left(\omega\varepsilon \frac{\partial E_z}{\partial x} + \beta \frac{\partial H_z}{\partial y}\right) \bigg/ (k^2 - \beta^2), \tag{7.12}$$

$$E_x = -j\left(\beta \frac{\partial E_z}{\partial x} + \omega\mu \frac{\partial H_z}{\partial y}\right) \bigg/ (k^2 - \beta^2), \tag{7.13}$$

$$E_y = j\left(-\beta \frac{\partial E_z}{\partial y} + \omega\mu \frac{\partial H_z}{\partial y}\right) \bigg/ (k^2 - \beta^2), \tag{7.14}$$

where

$$k^2 = \omega^2\mu\varepsilon. \tag{7.15}$$

Evidently, if E_z and H_z can be determined, the other components of $\mathbf{E}$ and $\mathbf{H}$ follow immediately. The basic wave equations governing $\mathbf{E}$ and $\mathbf{H}$ are obtained by elimination from Eqs. (7.03) and (7.04). After some vector algebra and with the assumption of constant μ and ε it is found that

$$\nabla^2 \mathbf{E} = -k^2 \mathbf{E}, \tag{7.16}$$
$$\nabla^2 \mathbf{H} = -k^2 \mathbf{H}. \tag{7.17}$$

Noting the $\exp(-j\beta z)$-dependence of the fields the z-component equations of the above give

$$\nabla_{\mathrm{T}}^2 E_z + (k^2 - \beta^2) E_z = 0 \tag{7.18}$$

and

$$\nabla_{\mathrm{T}}^2 H_z + (k^2 - \beta^2) H_z = 0, \tag{7.19}$$

where the transverse Laplacian operator

$$\nabla_{\mathrm{T}}^2 = \frac{\partial^2}{\partial x^2} + \frac{\partial^2}{\partial y^2} \tag{7.20}$$

is introduced.

Clearly, E_z and the transverse $\mathbf{E}$-field components determined by it from Eqs. (7.11)–(7.14) may exist separately from H_z and its corresponding transverse components. Thus two separate modes, transverse electric (TE) when $E_z = 0$, $H_z \neq 0$ and transverse magnetic (TM) when $H_z = 0$, $E_z \neq 0$ are identified. Equations (7.18) and (7.19) are crucial to this analysis, since all the other field quantities follow relatively trivially from E_z and H_z. It is noted that the equations take a simplified form of Eq. (3.02), the Helmholtz equation, so that a solution may be contemplated by variational methods. Before examining this approach in detail the boundary conditions holding in each of the two cases must be established. To a good approximation it is sufficient for most purposes to assume that the waveguide boundaries are perfect conductors. Where this is not good enough a more accurate solution taking wall losses into account may be obtained as a perturbation of the ideal-situation expressions. For a perfectly conducting waveguide the boundary rules of Section 1 require that E_z should vanish at the waveguide walls. Thus the constraint to be associated with Eq. (7.18) is simply the principal (Dirichlet) boundary condition $E_z = 0$. Since all the TM field components become determined by the Maxwell-derived equations, Eqs. (7.11)–(7.14), once E_z has been established subject to its boundary constraint, it follows that the further constraints required upon the other components of $\mathbf{E}$ and $\mathbf{H}$ will be satisfied automatically. However, turning to Eq. (7.19) the boundary rules of Section 1 give no direct clue as to what constraint should be applied to H_z. Suppose a portion of the boundary lies in the x–z-plane. Equations (7.12) and (7.13) give for the TE mode

$$H_y = \frac{-j\beta}{k^2 - \beta^2} \frac{\partial H_z}{\partial y}, \tag{7.21}$$

$$E_x = \frac{-j\omega\mu}{k^2 - \beta^2} \frac{\partial H_z}{\partial y}. \tag{7.22}$$

The explicit boundary rules require that the normal **H**-field and tangential **E**-field should vanish, in the case here of a boundary element in the x–z-plane, specifically H_y and E_x respectively. Both requirements are satisfied if $\partial H_z / \partial y = 0$ and it may be observed that $\partial/\partial y$ in this case is the *normal* derivative. Since the choice of the transverse (x, y) axes is arbitrary, it may safely be assumed that the boundary constraint upon H_z for the TE modes is $\partial H_z / \partial n = 0$ whatever the orientation of the boundary element. Evidently, Eq. (7.19) determining the TE modes is subject to the natural (homogeneous Neumann) boundary constraint. Although the differential equations governing TE and TM modes are the same, it is now seen that the distinct character of each of these modes is assured by their different boundary constraints.

7.2   *Variational formulation*

The variational formulation for uniform waveguide modes is now considered. By comparison of Eq. (7.18) with Eq. (3.02) it is seen that for TM modes the functional (3.06) reduces to

$$F(E_z) = \tfrac{1}{2} \int_\Omega [(\nabla E_z)^2 - (k^2 - \beta^2)E_z^2] \, \mathrm{d}x \, \mathrm{d}y, \tag{7.23}$$

subject to the Dirichlet boundary condition $E_z = 0$ on the waveguide boundary of the two-dimensional region Ω. A similar functional in H_z subject to the natural boundary condition $\partial H_z / \partial n = 0$ applies to the the TE mode. Writing $k_c^2 = k^2 - \beta^2$ it is recognised that Eqs. (7.18) and (7.19) represent *eigenvalue* problems with nontrivial solutions occurring only when k_c takes on one of a discrete set of values whose determination is part of the problem. The calculation of this set of numbers by finite element methods is developed in Chapter 3. Here it is merely noted that for each of the two mode types there is a complete sequence of distinct waves corresponding to the set of *cut-off wavenumbers* k_{ci} and *eigenfunctions* H_{zi}. Then with

$$\beta_i^2 = \omega^2 \mu\varepsilon - k_{ci}^2, \tag{7.24}$$

it is seen that each distinct mode has a different propagation constant and field pattern. It is observed that at any given frequency a particular eigenvalue k_{ci} may be large enough to require β_i^2 to be negative. In such a case the propagation constant $\gamma = \alpha + j\beta$ is purely real and the waveguide fields die away rapidly with z where previously they propagated as unattenuated waves. The waveguide fields are then described as *evanescent*. There exists for each mode i a lowest or *cut-off* frequency ω_{ci} corresponding to $\beta_i = 0$ with

$$\omega_{ci}^2 \mu\varepsilon = k_{ci}^2, \tag{7.25}$$

below which frequency the mode does not propagate in the ordinary sense but dies away exponentially with distance.

8.   Scalar Laplace and Helmholtz problems in three dimensions

The inhomogeneous Helmholtz equation, Eq. (3.02) and its corresponding functional (3.06) have so far only been discussed in the context of problems which have symmetries allowing them effectively to be treated in a two-dimensional form. When the dependent variable is a single scalar quantity there is no particular extra conceptual difficulty in dealing with truly three-dimensional problems. There is of course an extra and sometimes prodigious computational effort required to be applied in such cases when a problem is to be solved numerically to any accuracy. However, this latter aspect does not concern us in the present chapter. Here, the formulation of two practical problems which typify the approach using a three-dimensional scalar dependent variable are examined. Firstly, the computation of the resistance of an irregularly-shaped body is considered. An example might be the arrangement depicted in Fig. 2.13, a composite of two different materials having conductivities σ_1 and σ_2. Contact is made by means of two highly conductive electrodes maintained at potentials V_1 and V_2 whilst a current I flows. Clearly, the required resistance is

$$R = (V_2 - V_1)/I. \tag{8.01}$$

Fig. 2.13. Current flow through an irregularly-shaped body.

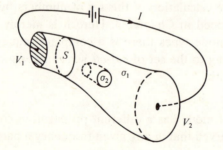

The material is characterised by Ohm's law in its vector point form

$$\mathbf{J} = \sigma \mathbf{E}. \tag{8.02}$$

Consider the steady state,

$$\mathbf{E} = -\nabla V, \tag{8.03}$$

so that

$$\mathbf{J} = -\sigma \nabla V. \tag{8.04}$$

With no time variation the equation of continuity for current, Eq. (1.08), requires the divergence of $\mathbf{J}$ to vanish. Thus, from Eq. (8.04),

$$\nabla \cdot (\sigma \nabla V) = 0. \tag{8.05}$$

Equation (8.05), subject to boundary constraints, defines a potential whose determination effectively solves the problem. The boundary conditions at the contact electrodes are clearly $V = V_1$ and $V = V_2$ respectively. On the free surface of the body it is observed that the normal component of current must vanish, since current cannot flow out into the surrounding insulating space. Through Eq. (8.04) it is seen that the normal component of ∇V, $\partial V / \partial n$ say, must consequently vanish. This is recognised as the homogeneous Neumann natural boundary condition. Equation (8.05) is a simplified version of the very general equation considered in Section 3, Eq. (3.02). Thus the correspondingly simplified functional (3.05)

$$F(U) = \tfrac{1}{2} \int_{\Omega} \sigma (\nabla U)^2 \, d\Omega \tag{8.06}$$

is stationary about $U = V$, true solution of Eq. (8.05), provided U is constrained to fit the Dirichlet boundary conditions which apply at the contact-electrode surfaces. Once again, it is significant to observe that no constraint is required on W with respect to the surfaces where the natural, Neumann condition applies. If σ is piecewise constant, as is suggested in the example of Fig. 2.13, Eq. (8.05) reduces to Laplace's equation $\nabla^2 V = 0$. However, in order to exploit the stationary nature of the functional, summed from the piecewise-constant conductivity regions, and to arrive at solutions which satisfy the interface conditions between such regions, the σ weighting when performing the integration of Eq. (8.06) clearly must be retained. Having determined the potential distribution V, the actual resistance required would be obtainable by performing the integration

$$I = - \int_{S} \sigma (\nabla V) \cdot d\mathbf{S} \tag{8.07}$$

over any cross-section S of the arrangement carrying the whole current, such as is illustrated in Fig. 2.13. The current so obtained may then be substituted into Eq. (8.01).

8.1   *An acoustic problem*

Secondly, a typical three-dimensional problem is now considered, departing from strictly electrical topics to consider acoustic resonance within a rigid air-filled enclosure. This problem is, nevertheless, an appropriate example since it illustrates the original use of the equation by Helmholtz, whose name it now bears. The application here is the calculation of acoustic resonance frequencies of halls, loudspeaker enclosures and so forth. The acoustic theory applicable to the problem is briefly set down, the reader being referred to standard works in the subject, such as that by Meyer & Neumann (1972), for instance, for detailed explanation. Sound waves in air are represented by longitudinal disturbances in pressure, p say, perturbations of a prevailing static pressure p_0. Such disturbances are governed by the wave equation

$$\nabla^2 p = \frac{1}{c^2} \frac{\partial^2 p}{\partial t^2}, \tag{8.08}$$

where c is the speed of sound in air

$$c = \gamma p_0 / \rho_0, \tag{8.09}$$

with ρ_0 the static density of the air and γ its ratio of specific heats at constant pressure and volume. The disturbance velocity of the air $\mathbf{v}$ is related to the incremental pressure p through a linearised equation of motion

$$\nabla p = -\rho_0 \frac{\partial \mathbf{v}}{\partial t}. \tag{8.10}$$

Equations (8.08) and (8.10) may be cast into complex phasor form to give

$$\nabla^2 p + k^2 p = 0, \tag{8.11}$$

$$\nabla p = -j\omega\rho_0 \mathbf{v}, \tag{8.12}$$

where

$$k^2 = \omega^2 / c^2. \tag{8.13}$$

It is recognised that Eq. (8.11) is in the Helmholtz form of Eq. (3.02) so that the results of Section 3 may be applied immediately provided appropriate boundary conditions are known. For rigid-walled containers Eq. (8.12) gives the required constraint on p, since at such walls the normal component of $\mathbf{v}$ must vanish. Evidently, the constraint is the

natural condition $\partial p/\partial n = 0$. From Eq. (3.06) the functional corresponding to Eq. (8.08) is

$$F(p) = \tfrac{1}{2} \int_{\Omega} [(\nabla p)^2 - k^2 p^2]\, d\Omega, \tag{8.14}$$

which is stationary to perturbations about correct solutions p. It will be
shown in Chapter 3 how making $F(p)$ stationary for a connected system
of finite elements yields a matrix equation whose eigenvalues represent
the sequence of values k_i which allow nontrivial solution of Eq. (8.11).
Since $k^2 = \omega^2/c^2$, this determination gives the required resonant
frequencies of the cavity. The problem here is also considered in its
three-dimensional form in Chapter 6.

9.   Readings

The task of specifying reading material adjunct to this chapter
is different from that corresponding to the other chapters of this text.
Chapter 2 is a review of electromagnetic field theory, included in an
attempt to indicate what problems may be considered for solution by
finite element methods, what are the variables concerned and what are
the laws obeyed. The subject of field theory is long established, so that
it is not necessary to quote original papers, whilst the texts listed are
but a selection from those available. The authors themselves, Silvester
(1968) and Ferrari (1975), have each written shortish works on electromagnetism, which have the merit of being consistent in style with the
present book. The works by Popović (1971) and Ramo, Whinnery &
Van Duzer (1965) suit the level of the text here admirably. At a similar
level the books by Bleaney & Bleaney (1976), Sander & Reed (1978)
and Hayt (1974) may also be recommended. Stratton (1941) is the
standard reference for a comprehensive exposition of electromagnetics
theory.

The reader may wish to consult a mathematical text covering points
relevant to finite element analysis. At an elementary level the book by
Riley (1974) covers the topics well, whilst the standard comprehensive
reference is the two-volume work by Morse & Feshbach (1953).

References

Bleaney, B. I. & Bleaney, B. (1976). *Electricity and Magnetism*, 3rd edn. Oxford:
    Clarendon.
Ferrari, R. L. (1975). *An Introduction to Electromagnetic Fields*. London: Van Nostrand
    Reinhold.

Hayt, W. H. (1974). *Engineering Electromagnetics*, 3rd edn. New York: McGraw-Hill.

Meyer, E. & Neumann, E. G. (1972). *Physical and Applied Acoustics: An Introduction.* New York: Academic Press.

Morse, P. N. & Feshbach, H. (1953). *Methods of Theoretical Physics*, vols. 1 and 2. New York: McGraw-Hill.

Popović, B. D. (1971). *Introductory Engineering Electromagnetics.* Reading, Mass.: Addison-Wesley.

Ramo, S., Whinnery, J. R. & Van Duzer, T. (1965). *Fields & Waves in Communications Electronics.* New York: Wiley.

Riley, K. F. (1974). *Mathematical Methods for the Physical Sciences.* Cambridge University Press.

Sander, K. F. & Reed, G. A. L. (1978). *Transmission and Propagation of Electromagnetic Waves*, Cambridge University Press.

Schelkunoff, S. A. (1943). *Electromagnetic Waves.* Princeton, N.J.: D. Van Nostrand.

Silvester, P. (1968). *Modern Electromagnetic Fields.* Englewood Cliffs: Prentice-Hall.

Stratton, J. A. (1941). *Electromagnetic Theory.* New York: McGraw-Hill.

3

Triangular elements for the scalar Helmholtz equation

1. Introduction

Any polygon, no matter how irregular, can be represented exactly as a union of triangles. It is thus reasonable to employ the triangle as the fundamental element shape, as in Chapter 1.

However, the solution accuracy can be raised markedly by using piecewise polynomials instead of piecewise-planar functions on each triangle. If desired, the increased accuracy can be traded for computing cost by using high-order approximations on each triangle but choosing much larger triangles than in the first-order method. Indeed, both theory and experience indicate that, for many two-dimensional problems, it is best to subdivide the problem region into the smallest possible number of large triangles, and to achieve the desired solution accuracy by the use of high-order polynomial approximations on this very coarse mesh.

In the following, details will be given for the construction of triangular elements for the inhomogeneous, scalar Helmholtz equation. Scalar and quasi-scalar problems will be considered throughout, while all materials will be assumed locally linear and isotropic. The interest in constructing elements for the inhomogeneous Helmholtz equation resides in its generality, which allows problems in Laplace's equation, Poisson's equation or the homogeneous Helmholtz equation to be solved also, by merely dropping terms from the general equation.

2. Simplex coordinates

The approximating functions used with first-order, triangular, finite elements possess two fundamental properties which account for the simplicity and attractiveness of the method. First, the functions can always be chosen to guarantee continuity of the desired potential or wave function across all boundaries between triangles, provided only that continuity is imposed at triangle vertices. Second, the approximations obtained are in no way dependent on the placement of the triangles with respect to the global x–y-coordinate system. The latter point is probably geometrically obvious: the solution surface is locally defined by the three vertex values of potential and is therefore unaltered by redefinition of the x- and y-axes, even though the algebraic expressions for the approximating functions on each triangle change. Both of these desirable properties are retained in high-order approximation, provided the local basis functions within each triangle are defined in the proper way.

The easiest approach to suitable basis polynomials is in terms of the so-called simplex coordinates. In general, a simplex in N-space is defined as the minimal possible nontrivial geometric figure in that space; it is always a figure defined by $N + 1$ vertices. Thus a one-dimensional simplex is a line segment, a two-dimensional simplex is a triangle, a simplex in three dimensions is a tetrahedron. Its vertex locations serve to define any simplex uniquely. The size of a simplex may be defined by

$$\sigma(S) = \frac{1}{N!} \begin{vmatrix} 1 & x_1^{(1)} & x_1^{(2)} \cdots & x_1^{(N)} \\ 1 & x_2^{(1)} & & \\ \vdots & & & \\ 1 & x_{N+1}^{(1)} & & x_{N+1}^{(N)} \end{vmatrix}. \tag{2.01}$$

The elements of this determinant are the vertex coordinates of the simplex. The subscripts identify the vertices, while the superscripts denote space directions. Under this definition, the size of a triangle is its area, the size of a tetrahedron its volume, and so on.

Let some point P be located inside the simplex S. Then P uniquely defines a subdivision of S into $N + 1$ subsimplexes, each of which has the point P as one vertex, the remaining N vertices being any N taken from the set of $N + 1$ vertices of S. For example, a line segment is partitioned into two line segments by a point P placed somewhere along its length; similarly, a triangle is partitioned into three subtriangles by any interior point, as shown in Fig. 3.1(a). Clearly, each of the $N + 1$ subsimplexes is entirely contained within the simplex S. The size of S must be the sum of the sizes of the subsimplexes,

$$\sigma(S) = \sum_{k=1}^{N+1} \sigma(S_k). \tag{2.02}$$

This relationship may be obtained geometrically by inspection, or it may be derived algebraically from the defining equation, Eq. (2.01).

The splitting of S into subsimplexes is completely defined by the choice of point P. Conversely, the sizes of the $N+1$ subsimplexes (relative to the size of S) may be regarded as defining the location of point P within the simplex S. Thus, one may define $N+1$ numbers,

$$\zeta_m = \sigma(S_m)/\sigma(S), \tag{2.03}$$

which specify the point P uniquely within S. These numbers are usually known as the *simplex coordinates* (or the homogeneous coordinates, or the barycentric coordinates) of P. Geometrically, they measure perpendicular distance toward one vertex from the opposite side or face, as indicated in Fig. 3.1(b). The reason will be evident, at least for triangles, if one observes that the ratio of areas of triangles 1–2–3 and P–2–3 in Fig. 3.1 is identical to the ratio of triangle heights, since both triangles have the line 2–3 as base. Similarly, it is geometrically evident as well as easy to deduce from Eqs. (2.02) and (2.03) that

$$\sum_{k=1}^{N+1} \zeta_k = 1. \tag{2.04}$$

This linear dependence of the simplex coordinates should not be surprising since it takes $N+1$ simplex coordinates to define a point in N-space.

Conversion from global Cartesian coordinates to simplex coordinates locally is readily accomplished by Eq. (2.03). For example, let a triangle

Fig. 3.1. (*a*) **Any interior point P defines a splitting of the triangle into subtriangles. (*b*) Simplex coordinates measure relative distance towards each vertex from the opposite side.**

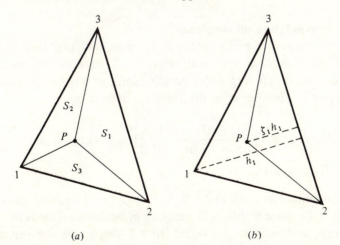

be placed in the x–y-plane. The first simplex coordinate of the point (x, y) is then given by

$$
\zeta_1 = \begin{vmatrix} 1 & x & y \\ 1 & x_2 & y_2 \\ 1 & x_3 & y_3 \end{vmatrix} \Bigg/ \begin{vmatrix} 1 & x_1 & y_1 \\ 1 & x_2 & y_2 \\ 1 & x_3 & y_3 \end{vmatrix}.
\tag{2.05}
$$

Expanding by minors of the first column, one obtains

$$
\zeta_1 = \frac{1}{2A} \left\{ \begin{vmatrix} x_2 & y_2 \\ x_3 & y_3 \end{vmatrix} - \begin{vmatrix} x & y \\ x_3 & y_3 \end{vmatrix} + \begin{vmatrix} x & y \\ x_2 & y_2 \end{vmatrix} \right\},
\tag{2.06}
$$

where A represents the triangle area. On rearranging, there results

$$
\zeta_1 = \frac{x_2 y_3 - x_3 y_2}{2A} + \frac{y_2 - y_3}{2A} x + \frac{x_3 - x_2}{2A} y.
\tag{2.07}
$$

A similar calculation for the remaining two simplex coordinates thus yields the conversion equation

$$
\begin{bmatrix} \zeta_1 \\ \zeta_2 \\ \zeta_3 \end{bmatrix} = \begin{bmatrix} x_2 y_3 - x_3 y_2 & y_2 - y_3 & x_3 - x_2 \\ x_3 y_1 - x_1 y_3 & y_3 - y_1 & x_1 - x_3 \\ x_1 y_2 - x_2 y_1 & y_1 - y_2 & x_2 - x_1 \end{bmatrix} \begin{bmatrix} 1 \\ x \\ y \end{bmatrix}.
\tag{2.08}
$$

It should be noted that the simplex coordinates are purely local in character; Eq. (2.03) yields the same values of simplex coordinates for a given point, regardless of any translation or rotation of the global coordinate system. This property is of great value. It allows most of the mathematical formulation of element equations to be accomplished once and for all, for a prototypal triangle; subsequently, the results can be applied to any triangle whatever by means of a few simple coordinate transformation rules.

3.   Interpolation on simplexes

To construct finite elements for potential problems, suitable interpolation functions are next required. Such polynomials can be readily defined, beginning with a certain family of auxiliary polynomials R of degree n. Member m of this family is defined by

$$
R_m(n, \zeta) = \prod_{k=0}^{m-1} \frac{\zeta - k/n}{m/n - k/n} = \frac{1}{m!} \prod_{k=0}^{m-1} (n\zeta - k), \quad m > 0
$$
$$
R_0(n, \zeta) = 1
\tag{3.01}
$$

This polynomial has zeros at $\zeta = 0, 1/n, \ldots, (m-1)/n$, and unity value at $\zeta = m/n$. In other words, it has exactly m equispaced zeros to the left of $\zeta = m/n$, and none to the right. On a 1-simplex (a line segment of some given length), a family of polynomials of degree n, with exactly n zeros each, may be defined by

$$\alpha_{ij} = R_i(n, \zeta_1)R_j(n, \zeta_2),\qquad(3.02)$$

where it is understood that $i + j = n$. The construction of the functions α_{ij} is indicated in Fig. 3.2(a). They may be recognised as being nothing more than the classical Lagrange interpolation polynomials, though expressed in an unusual notation. It may be noted that each interpolation polynomial, and each interpolation node, is denoted by a double index indicating its placement in the two simplex coordinates. This manner of labelling emphasises the relationship between interpolation functions and their corresponding interpolation points, and at the same time exhibits the symmetry inherent in a simplex. Fig. 3.2(b) shows the complete cubic family of interpolation polynomials constructed in this manner.

Fig. 3.2. (a) A line segment, and the polynomials $R_1(3, \zeta)$ and $R_2(3, \zeta)$, for which $i + j = 3$, but which do not vanish at node 21. (b) The four cubic interpolation polynomials for the line segment.

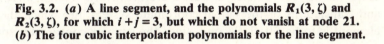

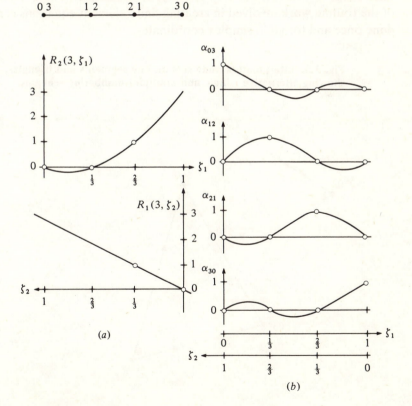

For triangles, exactly the same method is followed. Interpolation functions are defined by

$$\alpha_{ijk} = R_i(n, \zeta_1)R_j(n, \zeta_2)R_k(n, \zeta_3) \qquad i+j+k = n. \tag{3.03}$$

Again, the resulting polynomials are interpolatory on regularly spaced point sets, as shown in Fig. 3.3. It is easily seen that there are $(n+1) \times (n+2)/2$ such points on the triangle, and that the natural point numbering involves three indices. In many applications, the multi-index numbering becomes inconvenient because of the great length of index strings, and a single-index numbering scheme is preferred even though it does not clearly exhibit the geometric symmetry inherent in a simplex. Corresponding single-index and triple-index numbering schemes are also shown in Fig. 3.3.

The polynomials defined by Eqs. (3.02)–(3.03) possess two properties important in finite element construction. First, they are defined in terms of the simplex coordinates, and are therefore invariant under all rotations and translations of the global coordinate system into which the finite element is to be fitted. Second, since the simplex coordinates span the same range of values (from zero to unity) in any simplex whatever, most of the routine work involved in creating finite element equations can be done once and for all in simplex coordinates.

Fig. 3.3. Interpolation node sets on line segments and triangles, showing alternative (single and multiple) numbering schemes.

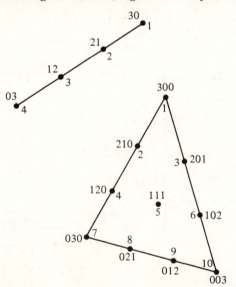

4.   Plane triangular elements

Various problems in electromagnetics may be regarded as special cases of the inhomogeneous Helmholtz equation

$$(\nabla^2 + k^2)u = g, \tag{4.01}$$

subject to various boundary conditions, of which the homogeneous Neumann condition (vanishing normal derivative) or Dirichlet conditions (fixed boundary value) are probably most commonly encountered in applications. As shown in detail in Chapter 2, solutions of Eq. (4.01) render stationary the functional

$$F(U) = \tfrac{1}{2} \int \nabla U \cdot \nabla U \, d\Omega - \tfrac{1}{2} k^2 \int U^2 \, d\Omega + \int Ug \, d\Omega, \tag{4.02}$$

provided the functions U are continuous within the problem region, and provided they satisfy the Dirichlet boundary conditions of the problem.

To construct an approximate solution by means of high-order triangular elements, a procedure will be used which resembles that followed for first-order elements. The problem region is approximated by a union of triangles, and on each triangle the potential function is represented as a linear combination of the approximating functions developed above,

$$U(\zeta_1, \zeta_2, \zeta_3) = \sum_{ijk} U_{ijk} \alpha_{ijk}(\zeta_1, \zeta_2, \zeta_3). \tag{4.03}$$

Each of these approximating functions vanishes at every nodal point, except at its associated node where it has unity value. Hence, the coefficients in (4.03) will be potential values at the interpolation nodes. Furthermore, interelement continuity will be guaranteed, as shown by the following argument. If $U(x, y)$ is a polynomial function of degree at most n in x and y jointly, then along any straight line in the x–y-plane the function $U = U(s)$ must correspondingly be a polynomial of at most degree n in the distance s measured along the straight line. A triangular finite element of order n has exactly $n + 1$ nodes along each triangle edge, Since U must be a polynomial function of degree n along each triangle edge, these $n + 1$ nodal values fix the value of U everywhere along the edge. Clearly, the edge nodes are invariably shared between every pair of adjacent triangles. Therefore, the potential variation on either side of the interelement boundary is determined by the same $n + 1$ parameters and U must be continuous at triangle edges.

A single triangular element will now be developed in detail, interconnection of elements to form the entire problem region being deferred until later. Substituting the approximation (4.03) into (4.02), the functional $F(U)$ becomes

$$F(U) = \tfrac{1}{2} \sum_i \sum_j U_i U_j \int \nabla \alpha_i \cdot \nabla \alpha_j \, d\Omega$$

$$- \frac{k^2}{2} \sum_i \sum_j U_i U_j \int \alpha_i \alpha_j \, d\Omega + \sum_i U_i \int \alpha_i g \, d\Omega. \tag{4.04}$$

Although not in principle necessary, the forcing function $g(x, y)$ is frequently approximated over each triangle by the same polynomials as the potential or wave functions U,

$$g = \sum_j g_j \alpha_j. \tag{4.05}$$

There are various practical ways of determining the coefficients in this approximation. For example, interpolative approximation may be used; the coefficients are then simply the values of $g(x, y)$ at the interpolation nodes. Using an approximation of the form (4.05), whatever the method for finding coefficients, the functional $F(U)$ may be expressed as the matrix form

$$F(U) = \tfrac{1}{2} U^T S U - \frac{k^2}{2} U^T T U + U^T T G. \tag{4.06}$$

Here, U is the vector of coefficients in (4.03), while G represents the vector of coefficients in (4.05); the square matrices S and T are given by

$$S_{mn} = \int \nabla \alpha_m \cdot \nabla \alpha_n \, d\Omega \tag{4.07}$$

and

$$T_{mn} = \int \alpha_m \alpha_n \, d\Omega. \tag{4.08}$$

The matrices S and T are sometimes termed the Dirichlet matrix, and the metric, associated with a particular set of approximating functions. Similarly derived matrices applicable to problems of elasticity are often referred to as the stiffness matrix and the mass matrix respectively.

The fundamental requirement is for $F(U)$ to be stationary. Since $F(U)$ in Eq. (4.06) is an ordinary function of a finite number of variables (the components of the array U), the stationarity requirement amounts to demanding that

$$\frac{\partial F}{\partial U_m} = 0 \tag{4.09}$$

for all m whose corresponding components of U are free to vary.

Substituting (4.06) into Eq. (4.09), and carrying out the indicated differentiations, there is obtained the matrix equation

$$\mathbf{SU} - k^2\mathbf{TU} = -\mathbf{T\dot{G}}. \tag{4.10}$$

Solution of this equation for $\mathbf{U}$ determines the approximate value of U everywhere in the region of interest, and thus solves the problem.

5.   High-order triangular-element matrices

The elements of the matrices $\mathbf{S}$ and $\mathbf{T}$ must be evaluated before proceeding further. To begin, the integrand of (4.01) may be written out in detail, yielding

$$S_{mn} = \int \left(\frac{\partial \alpha_m}{\partial x} \frac{\partial \alpha_n}{\partial x} + \frac{\partial \alpha_m}{\partial y} \frac{\partial \alpha_n}{\partial y} \right) d\Omega. \tag{5.01}$$

The chain rule of differentiation permits writing

$$\frac{\partial \alpha_m}{\partial x} = \sum_{i=1}^{3} \frac{\partial \alpha_m}{\partial \zeta_i} \frac{\partial \zeta_i}{\partial x}. \tag{5.02}$$

But, from (2.07) above, the typical derivative in (5.02) may be written

$$\frac{\partial \zeta_i}{\partial x} = \frac{y_{i+i} - y_{i-1}}{2A}. \tag{5.03}$$

On combining (5.02) and (5.03), and writing corresponding expressions for derivatives with respect to y, (5.01) can be recast entirely in terms of simplex coordinates for the triangle,

$$S_{mn} = \frac{1}{4A^2} \sum_i \sum_j (b_i b_j + c_i c_j) \int \frac{\partial \alpha_m}{\partial \zeta_i} \frac{\partial \alpha_n}{\partial \zeta_j} d\Omega. \tag{5.04}$$

Here, A represents the triangle area, while the geometric coefficients b and c are given by

$$b_i = y_{i+1} - y_{i-1}, \tag{5.05}$$

$$c_i = x_{i-1} - x_{i+1}. \tag{5.06}$$

Throughout the above discussion the subscripts always progress modulo 3, i.e., cyclically around the three vertices of the triangle. The double summations of Eq. (5.04) can be reduced to a single summation if it is recognised that

$$\left. \begin{array}{l} b_i b_j + c_i c_j = -2A \cot \theta_k \qquad i \neq j \\ b_i^2 + c_i^2 = 2A(\cot \theta_j + \cot \theta_k), \end{array} \right\} \tag{5.07}$$

where θ_i is the included angle at vertex i, and i, j, k denote the three vertices of the triangle. (This trigonometric identity may not be obvious; a proof is given in the Appendix.) Substituting (5.07) into Eq. (5.04), expanding, and collecting terms, the double summation may be replaced by a single summation,

$$S_{mn} = \sum_{k=1}^{3} \cot \theta_k \int \left(\frac{\partial \alpha_m}{\partial \zeta_{k+1}} - \frac{\partial \alpha_m}{\partial \zeta_{k-1}} \right) \left(\frac{\partial \alpha_n}{\partial \zeta_{k+1}} - \frac{\partial \alpha_n}{\partial \zeta_{k-1}} \right) \frac{d\Omega}{2A}. \tag{5.08}$$

The integral remaining on the right-hand side of this equation is dimensionless, and only involves quantities expressed in terms of the simplex coordinates. It may therefore be evaluated once and for all. Indeed, Eq. (5.08) may be written

$$S_{mn} = \sum_{k=1}^{3} Q_{mn}^{(k)} \cot \theta_k \tag{5.09}$$

to exhibit this fact more clearly. The three matrices **Q** are purely numeric, and have exactly the same values for a triangle of any size or shape whatever. They may therefore be calculated and tabulated for permanent reference. Details of the integration procedure required to evaluate (5.08) are lengthy, and will be found in the Appendix.

The metric **T** of Eq. (4.08) is even more easily calculated than **S**. Multiplying and dividing by the triangle area, (4.08) becomes

$$T_{mn} = 2A \int \alpha_m \alpha_n \frac{d\Omega}{2A}, \tag{5.10}$$

where the integral on the right-hand side is again dimensionless and independent of triangle size and shape, for it involves only quantities expressed in the simplex coordinates. It too may be evaluated once and for all. The use of simplex coordinates thus permits carrying out all differentiations and integrations in a universal manner, valid for any triangle. The triangle area ultimately appears as a multiplicative factor in the matrix **T**. Its shape, expressed by the cotangents of included angles at the three vertices, enters the matrix **S** as three weighting factors.

The integrations required in (5.08) and (5.10) are relatively easy. Writing the surface element of integration $d(\Omega/A)$ in terms of simplex coordinates, and integrating by parts,

$$\int \zeta_1^i \zeta_2^j \zeta_3^k \frac{d\Omega}{2A} \int_0^1 \int_0^{1-\zeta_1} \zeta_1^i \zeta_2^j (1 - \zeta_1 - \zeta_2)^k \; d\zeta_1 \, d\zeta_2. \tag{5.11}$$

it is shown in the Appendix that

$$\int \zeta_1^i \zeta_2^j \zeta_3^k \frac{d\Omega}{A} = \frac{i! \, j! \, k! \, 2!}{(i+j+k+2)!}. \tag{5.12}$$

Thus, the polynomial integrands can be multiplied out, and then integrated term-by-term. Since the results are of universal validity, element matrices up to and including order 6 have been tabulated. Table 3.1 gives the matrix values for the first three orders of elements.

6. Use of high-order triangular elements

The interconnection of elements to form a connected whole follows precisely the same methodology as in the first-order case, as set out in Chapter 1 (Section 4). For example, Fig. 3.4 shows the interconnection of two second-order triangular elements. This disjoint pair of elements possesses twelve independent nodal potentials, while the connected pair has only nine. The two potential vectors are related by the equation

$$\mathbf{U}_{dis} = \mathbf{C}\mathbf{U}_{con}, \tag{6.01}$$

where $\mathbf{C}$ is the connection matrix which expresses the constraints placed

Table 3.1. *Table of* $\mathbf{T}$ *and* $\mathbf{Q}$ *matrices for triangles*

$N = 1$
Common denominators: T 12
                        Q_1 2

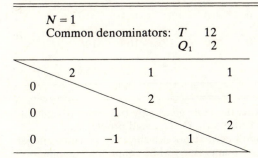

	2	1	1
0		2	1
0	1		2
0	−1	1	

$N = 2$
Common denominators: T 180
                        Q_1 6

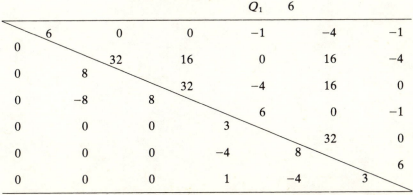

	6	0	0	−1	−4	−1
0		32	16	0	16	−4
0	8		32	−4	16	0
0	−8	8		6	0	−1
0	0	0	3		32	0
0	0	0	−4	8		6
0	0	0	1	−4	3	

Table 3.1 (*cont.*)

$$N = 3$$

Common denominators: T 6720
                                     Q_1 80

76	18	18	0	36	0	11	27	27	11
0	540	270	−189	162	−135	0	−135	−54	27
0	135	540	−135	162	−189	27	−54	−135	0
0	−135	135	540	162	−54	18	270	−135	27
0	−27	27	135	1944	162	36	162	162	36
0	0	0	−162	324	540	27	−135	270	18
0	27	−27	27	−162	135	76	18	0	11
0	3	−3	3	0	−3	34	540	−189	0
0	0	0	0	0	0	−54	135	540	18
0	0	0	0	0	0	27	−108	135	76
0	−3	3	−3			−7	27	−54	34

Since each **Q** and **T** are symmetric, the table gives only the upper triangle of **T**, and the lower triangle of the first matrix **Q**. To avoid problems of computer word-length and roundoff accumulation, values are given as integer quotients; the tabulated values are numerators, while common denominators are given for each individual tabulation. The elements of **T** given here are normalised with respect to $2A$ (see Eq. (5.10)). *Source:* after *International Journal of Engineering Science* (1969) **7**, 849–61.

on the set of twelve potentials. For the case shown in Fig. 3.4,

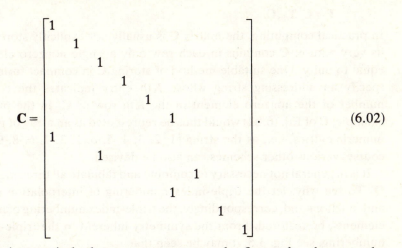

$$\mathbf{C} = \begin{bmatrix} 1 & & & & & & & & & & & \\ & 1 & & & & & & & & & & \\ & & 1 & & & & & & & & & \\ & & & 1 & & & & & & & & \\ & & & & 1 & & & & & & & \\ & & & & & 1 & & & & & & \\ 1 & & & & & & & & & & & \\ & 1 & & & & & & & & & & \\ & & & & & & & & 1 & & & \\ & & & & & & & & & 1 & & \\ & & & & & & & & & & 1 & \\ & & & & & & & & & & & 1 \end{bmatrix}. \tag{6.02}$$

Following precisely the same arguments as for first-order elements, the matrix **S** for the connected system is

$$\mathbf{S} = \mathbf{C}^{\mathrm{T}} \mathbf{S}_{\mathrm{dis}} \mathbf{C} \tag{6.03}$$

Fig. 3.4. Continuity of potential between adjacent elements is guaranteed by continuity of potential at the shared-edge nodes.

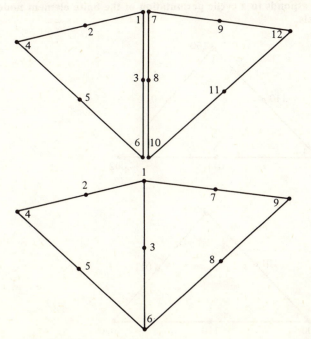

and the matrix **T** for the connected system is

$$\mathbf{T} = \mathbf{C}^{\mathrm{T}} \mathbf{T}_{\mathrm{dis}} \mathbf{C}. \qquad (6.04)$$

In practical computing, the matrix **C** is usually not explicitly stored. By its very nature, **C** contains in each row only a single nonzero element equal to unity. One suitable method of storing **C** in compact form is to specify an addressing string whose Kth entry indicates the column number of the nonzero element in the Kth row of **C**. In the present example, **C** of Eq. (6.02) would thus be represented as an array of twelve numeric entries, i.e., as the string [1, 2, 3, 4, 5, 6, 1, 3, 7, 6, 8, 9]. Of course, various other schemes can also be devised.

It is in general not necessary to compute and tabulate all three matrices **Q**. To see why, let the triple-index numbering of interpolation nodes and functions and, correspondingly, the triple-index numbering of matrix elements, be restored. From the symmetry inherent in the triple-index numbering, see Fig. 3.5, it may be seen that

$$\left. \frac{\partial \alpha_{ijk}}{\partial \zeta_1} \right|_{rst} = \left. \frac{\partial \alpha_{kij}}{\partial \zeta_2} \right|_{trs} = \left. \frac{\partial \alpha_{jki}}{\partial \zeta_3} \right|_{str}. \qquad (6.05)$$

In other words: all three directional derivatives, and therefore also all

Fig. 3.5. A cyclic permutation of the triangle coordinate labels corresponds to a cyclic permutation of the finite element node labels.

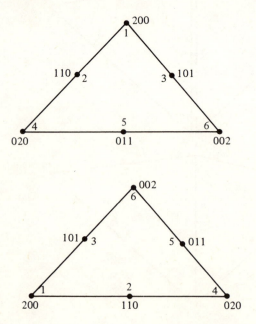

three matrices $\mathbf{Q}$, can be produced from a single one by permuting rows and columns in a manner which corresponds to advancing the triple indices cyclically. Each of the three matrices is therefore related to the other two by transformations of the form

$$\mathbf{Q}^{(2)} = \mathbf{PQ}^{(1)}\mathbf{P}^{\mathrm{T}}, \tag{6.06}$$

$$\mathbf{Q}^{(3)} = \mathbf{PQ}^{(2)}\mathbf{P}^{\mathrm{T}} = \mathbf{PPQ}^{(1)}\mathbf{P}^{\mathrm{T}}\mathbf{P}^{\mathrm{T}}. \tag{6.07}$$

Here, $\mathbf{P}$ represents the relevant permutation matrix.

The permutation transformations on the matrices $\mathbf{Q}$ have a straightforward geometric significance, as may be seen from Fig. 3.5. As defined above, $\mathbf{P}$ is a matrix whose columns are the columns of the unit matrix, rearranged to correspond to rotation of the triangle about its centroid in a counterclockwise direction. In Fig. 3.5, such a rotation carries point 6 into the location formerly occupied by point 1, point 3 into point 2, point 5 into point 3, etc. Correspondingly, the permutation matrix $\mathbf{P}$ is

$$\mathbf{P} = \begin{bmatrix} & & & 1 & & \\ & & & & 1 & \\ 1 & & & & & \\ & & & & & 1 \\ & 1 & & & & \\ & & 1 & & & \end{bmatrix}. \tag{6.08}$$

For brevity, it is often convenient to denote the permutation transformations by an operator, rot, which symbolises the subscript mappings required:

$$\mathbf{Q}^{(2)} = \mathrm{rot}\ \mathbf{Q}^{(1)}, \tag{6.09}$$

$$\mathbf{Q}^{(3)} = \mathrm{rot}\ \mathrm{rot}\ \mathbf{Q}^{(1)}. \tag{6.10}$$

Because the permutations represent rotations of the simplex coordinate numbering, it is hardly surprising that

$$\mathbf{Q}^{(1)} = \mathrm{rot}\ \mathrm{rot}\ \mathrm{rot}\ \mathbf{Q}^{(1)} \tag{6.11}$$

for three rotations of a triangle should map every point of the triangle into itself.

Any permutation matrix $\mathbf{P}$ consists of the columns of the unit matrix, so that only the location of the nonzero element in each column needs to be known to convey full information about it. Thus, the indexing string $IS = [6, 3, 5, 1, 2, 4]$ provides all the information required to perform matrix transformations of the form (6.09) on second-order elements. Indeed, the matrix transformation itself is best computationally executed, not as a multiplication but as a lookup operation in an indexing string, which substitutes one set of subscripts for another. For example, many Fortran language dialects will accept nested subscripts in the form S(IS(M), IS(N)).

Since the three matrices **Q** are easily derived from each other, only one such matrix needs to be computed and stored. Furthermore, since each **Q** is symmetric, only half its elements are actually required.

Turning next to the metric **T**, it is clear from Eq. (4.08) that this matrix is symmetric also. Practical computer programs therefore economise on memory by storing only one-half of **T**.

7.   High-order elements in waveguide analysis

As discussed in Chapter 2, the possible propagating modes of a perfectly conducting waveguide homogeneously filled with perfect dielectric are described by the wave functions u which satisfy

$$(\nabla^2 + k_c^2)u = 0, \qquad (7.01)$$

where k_c^2 is a cut-off wavenumber (see Section 7.2 of Chapter 2). This homogeneous Helmholtz equation is exactly of the form discussed above.

The finite element version of the problem becomes the solution of the matrix equation Eq. (4.10), but with zero right-hand side, subject to Dirichlet or Neuman boundary conditions for the TE and TM modes respectively. Here the primary object is to determine the unknown values k_c, which occur as an infinite set of *eigenvalues* when Eq. (7.01) itself is solved. The finite element discretisation Eq. (4.10) is treated by standard numerical techniques to yield a finite set of eigenvalues k which are approximations to the truncated set k_c. The eigensolutions U correspond to the waveguide field distribution, as explained in Chapter II Section 7.1.

High-order triangular elements may be expected to perform particularly well where the guide shape is a convex polygon, for the polygonal shape can be expressed exactly, as a union of triangles. Hence, the accuracy obtained from a finite element solution may be assessed by comparing with analytic, exact solutions, which are available for rectangular waveguides and for isosceles right-triangular guides. The rectangular shape may be decomposed into two or more triangles, while, for the triangular guide, quite accurate answers may be obtained even from a single element of high order!

Since triangular elements of the first six orders have been calculated and tabulated, the practical analyst will in each case need to decide whether to employ just a few high-order elements, or to choose a much larger number of elements of lower order. To gain an idea of error behaviour over a spectrum of modes, Table 3.2 shows the error encountered in calculating the cut-off wavelengths of the first eight TM modes of an isosceles right-triangular waveguide. Various orders of element were used, but all subdivisions were chosen in such a way as to keep the matrix equation exactly of order 55 each time. Hence, the required memory size and computing time were, for all practical purposes, the same in each case. Obviously, the use of few high-order elements produces much better results than a correspondingly larger number of low-order triangles.

Convex guide shapes do not have any reentrant corners at which any field singularities might occur. For such shapes, it can be shown on theoretical grounds that the error e behaves according to

$$e = Kh^{-2N}, \tag{7.02}$$

where h is the maximum dimension of any element in the model, and N is the polynomial order. K represents a proportionality constant, which is not in general known. It is evident from (7.02), as well as from Table 3.2, that the use of high-order polynomials is advantageous; what is gained from an increase in the exponent will always outweigh the loss from a corresponding increase in h.

For nonconvex guide shapes, where reentrant sharp corners are encountered, there does not exist an analytically solvable test case. However, comparisons with experimental measurements on real guides are possible. While the field singularities encountered in nonconvex shapes do produce a loss in accuracy, the use of high-order elements still remains attractive.

The approximating functions used in all the above are polynomial in each element, but only have continuity of the function sought (not of

Table 3.2. *Percentage error in calculated cut-off frequencies – right triangular waveguide*

Cut-off frequency	$N=1$ $E=144$	$N=2$ $E=36$	$N=3$ $E=16$	$N=4$ $E=9$
1.000 00	1.46	0.122	0.017	0.002
1.414 21	3.38	0.560	0.125	0.030
1.612 45	3.81	0.73	0.24	0.045
1.843 91	5.38	1.33	0.68	0.160
2.000 00	7.25	2.13	0.95	0.63
2.236 07	6.84	2.08	1.41	0.42
2.280 35	8.04	2.70	1.34	0.70
2.408 32	10.2	3.71	2.21	1.04

Note: N = order of elements, E = number of elements used in model.

its derivatives) across element interfaces. The resulting approximate solutions characteristically will show 'creases' at the element edges if the capabilities of the approximating functions are being stretched hard. For example, Fig. 3.6 shows two guided wave modes in a rectangular guide, calculated with two fifth-order elements. The fundamental TM mode of Fig. 3.6(a) is clearly very well approximated. The modal pattern in Fig. 3.6(b) should resemble a sequence of five patterns similar to Fig. 3.6(a) but, clearly, the fifth-order polynomials cannot quite manage to model so intricate a pattern accurately. Further, the diagonal interelement joint is clearly visible in Fig. 3.6(b).

8.   Axisymmetric scalar fields

Many scalar fields in isotropic but inhomogeneous media are described by the inhomogeneous Helmholtz equation

$$\nabla \cdot (p\nabla u) + k^2 u = g, \tag{8.01}$$

where p is the local value of some material property – permittivity, reluctivity, conductivity, etc. In many problems both the excitation function g and all boundary conditions possess rotational symmetry;

Fig. 3.6. Two modes of a rectangular waveguide, modelled by two fifth-order triangular elements. (a) The fundamental H (TM) mode. Although some error may be detected near the centre of the guide, the plot closely resembles the analytic solution. (b) The (5,1) H (TM) mode. In contrast to the fundamental mode, the element subdivision is quite clearly visible.

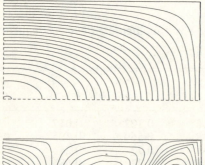

some examples are coaxial cable junctions, high-voltage insulators, and resonant cavities. In such cases the true mathematical problem is two-dimensional since finding the potential distribution over only one azimuthal plane suffices. The two-dimensional differential equation and the corresponding functional, however, differ from their translationally-uniform counterparts. New finite elements must therefore be devised.

As discussed in Chapter 2, such boundary-value problems are solved by minimising the two-dimensional (r–z-plane) functional

$$F(U) = \iint 2\pi r(p\nabla U \cdot \nabla U - k^2 U^2 + 2Ug)\, dr\, dz, \qquad (8.02)$$

where the differential and integral operations refer to the coordinates r and z only. This functional differs from its planar counterpart in having the integrand weighted by the radial variable r.

In the following, let R denote the solution region in the r–z-plane. Just as in the two-dimensional planar case, R will be subdivided into triangular elements, and the integral in (8.02) will be evaluated over each triangle in turn. As before, U and g will be approximated on each triangle, using the triangle interpolation polynomials. Thus

$$U = \sum_{m=1}^{n} U_m \alpha_m, \qquad (8.03)$$

$$g = \sum_{m=1}^{n} g_m \alpha_m. \qquad (8.04)$$

Element-independent integration is not yet possible because the factor r in the integrand is dependent on the position of the triangular element with respect to the axis of symmetry. The key to further progress lies in observing that r is trivially a linear function of the coordinates r and z, so that over any one element it may be expressed exactly as the linear interpolate of its three vertex values:

$$r = \sum_{i=1}^{3} r_i \zeta_i. \qquad (8.05)$$

Hence, the functional $F(U)$ may be written, substituting Eq. (8.05) into (8.02), as

$$F(U) = 2\pi \sum_{i=1}^{3} r_i \iint \zeta_i (p\nabla U \cdot \nabla U - k^2 U^2 + 2Ug)\, dr\, dz. \quad (8.06)$$

The derivatives with respect to r and z may be converted to derivatives with respect to the simplex coordinates, using exactly the same arguments as already advanced in the planar case. Thus,

$$\frac{\partial U}{\partial r} = \frac{1}{2A} \sum_{j=1}^{3} b_j \frac{\partial U}{\partial \zeta_j} \tag{8.07}$$

$$\frac{\partial U}{\partial z} = \frac{1}{2A} \sum_{j=1}^{3} c_j \frac{\partial U}{\partial \zeta_j}, \tag{8.08}$$

where

$$b_i = z_{i+1} - z_{i-1}, \tag{8.09}$$

$$c_i = r_{i-1} - r_{i+1}, \tag{8.10}$$

the subscript i being understood to progress cyclically around the triangle vertices, while A represents the triangle area. Substitution into (8.06), followed by rewriting the summations using the cotangent identity (5.07), then yields

$$F(U) = 2\pi \sum_{i=1}^{3} r_i \iint \zeta_i \left[\frac{p}{2A} \sum_{j=1}^{3} \cot \theta_j \left(\frac{\partial U}{\partial \zeta_{j-1}} - \frac{\partial U}{\partial \zeta_{j+1}} \right)^2 \right.$$
$$\left. - k^2 U^2 + 2Ug \right] dr\, dz. \tag{8.11}$$

The polynomial approximations (8.08) and (8.09) may next be substituted into Eq. (8.11). The stationarity condition for the functional F,

$$\frac{\partial F}{\partial U_i} = 0, \tag{8.12}$$

then reads, fully stated,

$$\sum_{i=1}^{3} r_i \left\{ \frac{p}{A} \sum_{m=1}^{n} \sum_{l=1}^{3} Q_{mk}^{(il)} \cot \theta_l U_m \right.$$
$$\left. - 2k^2 \sum_{m=1}^{n} R_{mk}^{(i)} U_m + 2 \sum_{m=1}^{n} R_{mk}^{(i)} g_m \right\} = 0, \tag{8.13}$$

where

$$Q_{mk}^{(il)} = \iint \zeta_i \left(\frac{\partial \alpha_k}{\partial \zeta_{l+1}} - \frac{\partial \alpha_k}{\partial \zeta_{l-1}} \right) \left(\frac{\partial \alpha_m}{\partial \zeta_{l+1}} - \frac{\partial \alpha_m}{\partial \zeta_{l-1}} \right) dr\, dz \tag{8.14}$$

and

$$R_{mk}^{(i)} = \iint \zeta_i \alpha_m \alpha_k \, dr\, sz. \tag{8.15}$$

Examination shows that all the above integrals are expressed entirely in terms of the triangle interpolation polynomials and the simplex coordinates. Consequently, the integrals are now all in a form in which they can be evaluated once and for all. The triangle size, shape, and placement relative to the axis of symmetry are of course left to be specified when the elements are actually assembled. In terms of the integral values, which are numerically defined and independent of triangle geometry and location, Eq. (8.13) may now be written in the form

$$\mathbf{SU} - k^2 \mathbf{TU} + \mathbf{TG} = 0, \tag{8.16}$$

which exactly accords with the form of matrix equation obtained for planar problems, provided the matrices $\mathbf{S}$ and $\mathbf{T}$ are defined by

$$\mathbf{S} = \frac{p}{A} \sum_i \sum_l \cot \theta_l Q^{(il)} r_i \tag{8.17}$$

and

$$\mathbf{T} = \sum_i r_i R^{(i)}. \tag{8.18}$$

In the axisymmetric case, placement of the triangle relative to the global coordinate system is of importance, while in the two-dimensional planar case, orientation with respect to the coordinate axes is irrelevant. This difference is clearly reflected in Eqs. (8.17) and (8.18), where the location and orientation of the triangle relative to the global coordinate system are explicitly specified by the radial coordinate values at the triangle vertices. It might be noted, on the other hand, that translation of a triangle in the z-direction does not alter the matrices. This is quite as it should be; in an axisymmetric problem, the choice of origin is arbitrary in the z-direction, while $r = 0$ must necessarily be placed at the axis of symmetry.

An interesting extension to the range of usefulness of the axisymmetric formulation may be perceived on examining Eq. (8.05). Since the three simplex coordinates sum to unity, one must have

$$\sum_{i=1}^{3} \iint \zeta_i f \, dr \, dz = \iint f \, dr \, dz. \tag{8.19}$$

Comparing with Eqs. (8.14) and (8.15), which define the arrays $\mathbf{Q}$ and $\mathbf{R}$, it is noted that summing over the index i collapses the nine axisymmetric matrices $\mathbf{Q}$ into the three corresponding matrices of translationally-symmetric analysis. Similarly, summation of the three arrays $\mathbf{R}$ of the axisymmetric problem over the index i yields exactly the matrix $\mathbf{T}$, Eq. (4.08). This fact allows matrices for the planar problem to be derived from those for the axisymmetric scalar case.

Among the nine distinct arrays $\mathbf{Q}$ and three arrays $\mathbf{R}$, there exist numerous symmetry properties which imply that only a few of these matrices are really independent. The general principles involved are similar to those applicable to the planar case, but the details are a little subtler. Three matrices $\mathbf{Q}$, and one $\mathbf{R}$, turn out to be independent, and need to be stored in computer programs.

9.   Solution of coaxial-line problems

The axisymmetric scalar elements are, clearly, ideally suited to the analysis of coaxial-line discontinuity problems. At the same time, coaxial-line problems with known solutions can provide some indication of the performance of these elements. The simplest possible problem of this class is that of an infinite coaxial line, readily modelled as a rectangular region in the r–z-plane with Neumann boundary conditions imposed on two sides parallel to the r-axis, Dirichlet boundary conditions on the remaining two. The rectangular region is easily subdivided into various numbers of triangles, permitting a rough assessment to be obtained of the relative performance of the various orders of elements. The analytic solution in this case is of course known, so that no difficulty attaches to comparing computed and exact results. Table 3.3 shows the radial variation of potential values obtained for six different element subdivisions and orders N.

As may be seen, the first-order solutions shown do not quite achieve two-significant-figure accuracy locally, while the second-order solution

Table 3.3. *Nodal potential values for a long coaxial structure*

r	$N = 1$	$N = 2$	$N = 3$	$N = 4$	$N = 5$	$N = 6$	exact
1.0	1.0000	1.0000	1.0000	1.0000	1.0000	1.0000	1.0000
2.0	0.7363	0.7334	0.7305	0.7283		0.7293	0.7298
2.2	0.7076				0.6922		0.6926
3.0	0.5782	0.5737	0.5716	0.5718		0.5717	0.5717
4.0	0.4652	0.4611	0.4593	0.4606		0.4593	0.4595
4.6	0.3948				0.4052		0.4050
5.0	0.3773	0.3736	0.3727	0.3723		0.3723	0.3723
5.8	0.3153				0.3144		0.3147
6.0	0.3053	0.3023	0.3017	0.3015		0.3015	0.3014
7.0	0.2438				0.2411		0.2413
8.0	0.1918	0.1898	0.1894	0.1893		0.1892	0.1893
8.2	0.1820				0.1797		0.1797
9.0	0.1453	0.1438	0.1435	0.1434		0.1433	0.1434
9.4	0.1282				0.1264		0.1264
10.0	0.1037	0.1026	0.1024	0.1023		0.1022	0.1023
10.6	0.0807				0.0796		0.0796
11.0	0.0660	0.0653	0.0652	0.0651		0.0651	0.0651
11.9	0.0383				0.0378		0.0378
12.0	0.0316	0.0313	0.0312	0.0312		0.0312	0.0312
13.0	0.0000	0.0000	0.0000	0.0000	0.0000	0.0000	0.0000
Number of elements							
	288	72	32	18	8	8	

deviates from the analytic result only in the third digit. The third-order solution provides full three-figure accuracy, while the fourth-order solution differs from the exact results only slightly in the fourth figure. Using elements of fifth order, a solution accuracy similar to fourth-order elements is observed, but the error behaviour has become erratic. Using sixth-order elements, this erratic behaviour is still more pronounced, showing that discretisation error has entirely vanished from the solution, while roundoff error due to the finite word-length used in computation has taken over as the governing factor in limiting accuracy. In the cases shown, matrix orders were roughly similar, ranging from 150 to 175. For this numerical experiment, a computer with 24-bit floating point mantissa was used, corresponding to approximately seven decimal digits. In matrix problems of this general character, four-figure accuracy may often be expected from seven-digit calculations. There are two principal sources for the roundoff error: error accumulation in the numerous additive processes involved in setting up the matrices, and error accumulation in the solution of simultaneous equations. Roundoff error accruing from the latter can be practically eliminated by means of iterative refinement in the equation-solving process. However, little can be done about error arising in the process of creating the equations, aside from performing all calculations in double precision.

10. Axisymmetric vector fields

Many field problems are not conveniently described by scalar potentials. In these cases, it becomes necessary to solve for a vector quantity, which may be one of the field vectors $\mathbf{E}$, $\mathbf{H}$, or the flux densities $\mathbf{D}$, $\mathbf{B}$; or it may be a vector potential. In any case, it is necessary then to solve a vector Helmholtz equation of the form

$$(\nabla^2 + k^2)\mathbf{A} = \mathbf{B}, \tag{10.01}$$

where $\mathbf{A}$ denotes the vector to be determined, while $\mathbf{B}$ is some known source vector. It should be noted that the Laplacian operator in (10.01) is the vector Laplacian,

$$\nabla^2 \mathbf{A} = \nabla \cdot \nabla \mathbf{A} - \nabla \times \nabla \times \mathbf{A}, \tag{10.02}$$

which is in general different from its scalar counterpart. The functional

$$F(\mathbf{U}) = \int (\nabla \times \mathbf{U} \cdot \nabla \times \mathbf{U} - \mathbf{U} \cdot \nabla \nabla \cdot \mathbf{U} - k^2 \mathbf{U} \cdot \mathbf{U}$$

$$+ 2\mathbf{B} \cdot \mathbf{U}) \, d\Omega - \oint \mathbf{U} \times \nabla \times \mathbf{U} \cdot d\mathbf{S} \tag{10.03}$$

is then minimised by the vector $\mathbf{U}$ nearest the correct solution $\mathbf{A}$ to

(10.01), provided that

$$\nabla\nabla \cdot \mathbf{U} = 0. \tag{10.04}$$

As in the scalar case, the most common types of boundary conditions that occur in practice cause the surface-integral term to vanish,

$$\oint \mathbf{U} \times \nabla \times \mathbf{U} \cdot d\mathbf{S} = 0. \tag{10.05}$$

For the moment, only axisymmetric situations will be considered in which both **B** and **A** have a single component, and that in the azimuthal direction. That is to say, all possible approximate solution vectors must be of the form

$$\mathbf{U} = U_\phi \hat{\boldsymbol{\phi}}, \tag{10.06}$$

where $\hat{\boldsymbol{\phi}}$ is the unit vector in the azimuthal direction. Under this restriction, the boundary integral of (10.05) becomes

$$\oint U_\phi \left(\frac{\partial U_\phi}{\partial n} + \frac{U_\phi}{r} \frac{dr}{dn} \right) dS = 0. \tag{10.07}$$

It will be seen that the homogeneous Dirichlet condition of vanishing **U** at surfaces satisfies the surface-integral requirement; so do various other commonly encountered situations.

Under the assumptions and restrictions outlined, the functional of Eq. (10.03) becomes finally

$$F(\mathbf{U}) = \int \left\{ r \left[\left(\frac{\partial U_\phi}{\partial z} \right)^2 + \left(\frac{\partial U_\phi}{\partial r} \right)^2 - k^2 U_\phi^2 + 2B_\phi U_\phi \right] \right.$$
$$\left. + 2U_\phi \frac{\partial U_\phi}{\partial r} + \frac{U_\phi^2}{r} \right\} d\Omega, \tag{10.08}$$

where the differentiations and integrations refer to the r–z-plane only. This functional is not only different from the scalar axisymmetric one; it is particularly inconvenient because it contains the singular term $1/r$, which does not admit a polynomial expansion. However, it can be brought into the general finite element framework by a change in variable: instead of solving for the vector originally desired, one solves for

$$U' = U_\phi / \sqrt{r}. \tag{10.09}$$

Correspondingly, it is convenient to write the right-hand term as

$$B' = B_\phi / \sqrt{r}, \tag{10.10}$$

so that the functional of (10.03) finally becomes

$$F(U') = \int r^2 \left[\left(\frac{\partial U'}{\partial r} \right)^2 + \left(\frac{\partial U'}{\partial z} \right)^2 \right] dR + \tfrac{9}{4} \int U'^2 \, dR$$

$$- k^2 \int r^2 U'^2 \, dR + 2 \int r^2 B' U' \, dR + \int r U' \frac{\partial U'}{\partial r} \, dR, \qquad (10.11)$$

where the region R of integration is again the relevant portion of the r–z-plane.

To develop finite elements for the axisymmetric vector problem, it remains to substitute the usual polynomial expansion for the trial function,

$$U' = \sum_j U'_j \alpha_j(\zeta_1, \zeta_2, \zeta_3). \qquad (10.12)$$

The expansion is lengthy because of the many terms in the functional (10.11), and it will not be carried out in detail here. However, it should be evident that all terms in (10.11) are now polynomials, and therefore directly integrable without approximation. Furthermore, the integrals are all expressible in terms of the simplex coordinates, so that once again the integrals can be calculated for a reference triangle whose shape, size, and placement eventually appear as weighting coefficients. The detailed generation of elements, as well as subsequent solution techniques, therefore follow exactly the same pattern as for the scalar field case.

11.   Readings

High-order finite elements are in widespread use in almost all branches of engineering analysis, having first been introduced in elasticity problems. A good source for background material in this area is the textbook by Desai & Abel (1972).

Systematic methods for generating interpolation functions on triangles were proposed at a relatively early date by Silvester (1969a), with immediate application to the solution of potential problems.

Electrical engineering applications of high-order simplex elements have been very varied. An early application was to the solution of waveguide problems (Silvester, 1969b). Of considerable interest is their use in diffusion problems by Kaper, Leaf & Lindeman (1972), who discuss high-order triangles extensively. A large-scale program using elements up to sixth order, with extensive application-oriented further processing of the field results, is reported by Kisak & Silvester (1975). The latter deals with complex diffusion. McAulay (1975) extends the technique to wave propagation in lossy media, while Stone (1973) and others have constructed matrices for the acoustic surface-wave problem. Straightforward potential problem solutions have been reported, for example, by Andersen (1973), who describes a large-scale applications package with extensive input and output data management facilities.

A large part of the work on simplex elements has led to the creation of tabulated universal matrices. Thus, in addition to the planar triangles already referred to, matrices for the axisymmetric cases (Konrad & Silvester, 1973) have been recorded in the literature. The general approach, construction of any 'universal' matrices from a few primitives, was given by Silvester (1978) more recently – no doubt another case of clear scientific hindsight!

References

Andersen, O. W. (1973). 'Transformer leakage flux program based on finite element method', *Institute of Electrical and Electronics Engineers Transactions on Power Apparatus and Systems*, **PAS–92**, 682–9.

Desai, C. S. & Abel, J. F. (1972). *Introduction to the Finite Element Method*. New York: Van Nostrand.

Kaper, H. G., Leaf, G. K. & Lindeman, A. J. (1972). 'The use of interpolatory polynomials for a finite element solution of the multigroup diffusion equation'. *In: The Mathematical Foundations of the Finite Element Method with Applications to Partial Differential Equations* (A. K. Aziz, ed.). New York: Academic Press, pp. 785–9. (*See also: Transactions of the American Nuclear Society*, **15**, 298.)

Kisak, E. & Silvester, P. (1975). 'A finite element program package for magnetotelluric modelling', *Computer Physics Communications*, **10**, 421–33.

Konrad, A. & Silvester, P. (1973). 'Triangular finite elements for the generalized Bessel equation of order m', *International Journal for Numerical Methods in Engineering*, **7**, 43–55.

McAulay, A. D. (1975). 'Track-guided radar for rapid transit headway control', *Journal of Aircraft*, **12**, 676–81.

Silvester, P. (1969a). 'High-order polynomial triangular finite elements for potential problems', *International Journal of Engineering Science*, **7**, 849–61.

Silvester, P. (1969b). 'A general high-order finite-element waveguide analysis program', *Institute of Electrical and Electronics Engineers Transactions on Microwave Theory and Techniques*, **MTT–21**, 538–42.

Silvester, P. (1978). 'Construction of triangular finite element universal matrices', *International Journal for Numerical Methods in Engineering*, **12**, 237–44.

Stone, G. O. (1973). 'High-order finite elements for inhomogeneous acoustic guiding structures', *Institute of Electrical and Electronics Engineers Transactions on Microwave Theory and Techniques*, **MTT–21**, 538–42.

4

Finite elements for integral operators

1. Introduction

The equation for the static scalar potential at distance r from a point charge q in a uniform medium of permittivity ε is

$$V = q/(4\pi\varepsilon r). \tag{1.01}$$

This relation gives a lead to an important approach, alternative to the partial differential equation method which has been applied so far to the solution of problems in electromagnetics. If it is supposed that a volume charge density ρ within a given volume Ω and a surface charge density σ over a given surface S are the sources of a static electric field, then from Eq. (1.01) the scalar potential due to such sources is

$$V(\mathbf{r}) = \int_{\Omega} \frac{\rho(\mathbf{r}') \, d\Omega'}{4\pi\varepsilon |\mathbf{r} - \mathbf{r}'|} + \int_{S} \frac{\sigma(\mathbf{r}') \, dS'}{4\pi\varepsilon |\mathbf{r} - \mathbf{r}'|}. \tag{1.02}$$

The notation in Eq. (1.02) refers to an element of charge at Q, position vector $\mathbf{r}'$, producing a field which is evaluated at the point P, position vector $\mathbf{r}$, as shown in Fig. 4.1. The position of the common origin O of $\mathbf{r}$ and $\mathbf{r}'$ is arbitrary and unimportant. Note that $\mathbf{r} - \mathbf{r}'$ corresponds to the vector $\mathbf{r}_1$ used in Chapter 2. The integrations are taken over the volume Ω and surface S appropriate to the charge distributions and with respect to the dummy variable $\mathbf{r}'$. Usually this is practical for finite volumes.

It has been observed earlier that a solution of the form (1.02) does in fact satisfy Maxwell's equations through the Poisson equation in potential, Eq. (2.06), Chapter 2. However, electrostatics problems are not ordinarily posed by specifying charge distributions. Rather, the potentials of conducting electrodes are given, the surface charge σ on such electrodes being regarded as an unknown along with the potential distribution $V(\mathbf{r})$ in the interelectrode space. In this case, supposing for simplicity that volume charge is absent, Eq. (1.02) can be written as an *integral equation* for the unknown σ:

$$V_0(\mathbf{r}) = \int_S \frac{\sigma(\mathbf{r}')\, dS'}{4\pi\varepsilon |\mathbf{r} - \mathbf{r}'|}, \tag{1.03}$$

with the points $\mathbf{r}$ and $\mathbf{r}'$ both lying on S and given the function $V_0(\mathbf{r})$ on S. Normally, V_0 will be piecewise constant, representing a system of electrodes charged to known potentials. On physical grounds it may be safely assumed that Eq. (1.03) does in fact uniquely determine σ. Then the unknown fields in the interelectrode space may be worked out using Eq. (1.02).

The function

$$G(\mathbf{r}|\mathbf{r}') = \frac{1}{4\pi\varepsilon |\mathbf{r} - \mathbf{r}'|}, \tag{1.04}$$

which appears in Eqs. (1.02) and (1.03) and is physically the potential at one place due to unit point charge at another, is described as a *Green's function*, in this case corresponding to an unbounded uniform medium. It should be noted that the Green's function (1.04) is singular at $\mathbf{r}' = \mathbf{r}$, typically the case. Nevertheless, it is evident on physical grounds that the integral (1.03) does exist – the singularity is *integrable*. Obviously, special care has to be exercised in numerical schemes which exploit integrals such as in Eq. (1.03).

In general it may be said that any electromagnetics problem described by a partial differential equation, with or without source terms and subject to prescribed boundary conditions, can alternatively be posed in integral equation form. Such formulation requires the finding of an appropriate Green's function. These are unfortunately problem dependent but, nevertheless, a number of cases are discussed in this chapter where the Green's function is known. The advantages to be gained in

Fig. 4.1. Source and field points for integral operations in electrostatics.

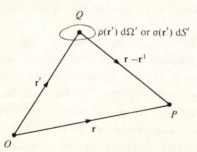

numerical schemes exploiting integral operators are principally a reduction in the number of variables which need to be handled in matrix operations and the elimination of the necessity to extend models to the infinite boundaries of open systems. The reduction in number of variables is offset to some extent by the fact that each finite element is usually related to every other, so that the resulting matrices are dense.

2. One-dimensional finite elements for the integral equations of electrostatics

In this section the problem of a coaxial line with arbitrary, constant cross-section is examined as an example illustrating a simple application of integral equation methods in finite element analysis. Subsequently, the particular case with rectangular geometry, previously considered in Chapter 2, Section 5, is reexamined.

For a translationally-symmetric, two-dimensional system, such as applying to the coaxial line, Eq. (1.01) may be replaced by

$$V = -q_l \frac{\ln (r/r_0)}{2\pi\varepsilon}. \tag{2.01}$$

In Eq. (2.01) q_l is the charge density per unit length of a line source extending to infinity in both directions. An apparently arbitrary radius $r = r_0$ is introduced to establish the reference zero of potential. When considering the three-dimensional, point-source case this reference was tacitly chosen to be at infinity and no arbitrary constant was needed. Here, however, it is noted that V from Eq. (2.01) becomes infinite as r increases without limit. Such nonphysical behaviour corresponds to the practical impossibility of realising the truly two-dimensional model, extending as it does to infinity in the axial direction.

Suppose there is no free charge in the space between the two coaxial-line conductors. Then the integral equation determining σ, the surface charge density on a surface S maintained at potential $V_0(\mathbf{r})$, is

$$V_0(\mathbf{r}) = -\frac{1}{2\pi\varepsilon} \int_S \sigma(\mathbf{r}') \ln \left| \frac{\mathbf{r}-\mathbf{r}'}{r_0} \right| ds', \tag{2.02}$$

where $\mathbf{r}$ and $\mathbf{r}'$ both are position vectors of points on the surface S. It is supposed that a unit length of the infinitely long coaxial system is being considered, so that ds' may be taken to be an elementary arc segment belonging to a two-dimensional curve S. Equation (2.02) can be applied directly to the geometry of Fig. 4.2 if the curve S is identified as the 'sum' of S_A, the coaxial inner conductor and S_B, its outer conductor. The potentials of S_A and S_B with respect to some common datum are taken as V_A and V_B respectively.

2.1   *Finite element solution of the integral equation*

Let the perimeter of the two conductors be divided into a total number altogether of M segments. The segments are numbered $i = 1, 2, \ldots, M$ and each has length Δs_i, not necessarily the same for all cases. It may be supposed that along each segment the surface charge density σ is represented by some function $\sigma_i(\mathbf{r})$. Each function $\sigma_i(\mathbf{r})$ would itself be regarded as being zero outside its subdomain Δs_i. Then with the Green's function here being expressed by

$$G(\mathbf{r}|\mathbf{r}') = -\frac{1}{2\pi\varepsilon} \ln \left| \frac{\mathbf{r} - \mathbf{r}'}{r_0} \right|, \tag{2.03}$$

it is seen that Eq. (2.02) may be written

$$V_0(\mathbf{r}) = \sum_{i=1}^{M} \int_{\Delta s_i} \sigma_i(\mathbf{r}') G(\mathbf{r}|\mathbf{r}') \, ds'. \tag{2.04}$$

Each function $\sigma_i(\mathbf{r}')$ is now approximated within its subdomain by a set of linearly independent *basis functions* p_{ik}:

$$\sigma_i(\mathbf{r}') = \sum_{k=1}^{n} L_{ik} p_{ik}(\mathbf{r}'). \tag{2.05}$$

Galerkin's method (also discussed in Chapter 6, Section 3.4) may now

Fig. 4.2. Coaxial line of arbitrary cross-section. The segments labelled i and j are shown as being on different members of the coaxial pair of conductors, but they may also be placed on the same member.

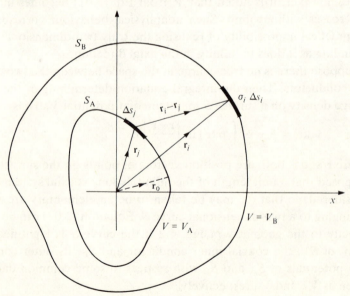

be applied to determine the coefficients L_{ik} and hence the charge distribution σ which is the solution to the problem. In the Galerkin procedure, Eq. (2.05) is substituted into Eq. (2.04). The resulting equation is multiplied both sides by $p_{jl}(\mathbf{r})$, one of the basis functions associated with the jth subdomain, Δs_j, and integrated over this subdomain giving

$$\int_{\Delta s_j} V_0(\mathbf{r}')p_{jl}(\mathbf{r}) \, ds = \sum_{i=1}^{M} \sum_{k=1}^{n} L_{ik} \int_{\Delta s_j} \int_{\Delta s_i} p_{ik}(\mathbf{r}')p_{jl}(\mathbf{r})$$

$$\times G(\mathbf{r}|\mathbf{r}') \, ds' \, ds \quad j = 1, 2, \dots, M. \tag{2.06}$$

The left-hand side of Eq. (2.06) represents a known vector of Mn terms. On the right-hand side of the equation it is observed that the double integral, although perhaps exceedingly difficult to evaluate, is nevertheless completely determinate. Thus Eq. (2.06) may be expressed in matrix form as

$$\mathbf{V} = \mathbf{PL}, \tag{2.07}$$

where $\mathbf{P}$ is an $Mn \times Mn$ square matrix of known coefficients whilst $\mathbf{L}$ is an unknown row vector of Mn terms which may be solved for from Eq. (2.07) by standard matrix techniques. The vector $\mathbf{L}$, once known, enables the charge distribution σ to be reconstructed via Eq. (2.05) and hence, for instance, the capacitance C per unit length of the coaxial system of Fig. 4.2 to be determined. It should be noted that C would be determined from the total charge on one of the plates, say

$$Q_A = \sum_{i=1}^{M_A} \int_{\Delta s_i} \sigma_i(\mathbf{r}) \, ds \tag{2.08}$$

and the relationship

$$Q_A = C(V_B - V_A). \tag{2.09}$$

Thus it is observed that, having obtained much detailed information in the vector $\mathbf{L}$, the operations required to determine the practical coefficient C average out a great deal of that fine detail.

2.2   *The piecewise-constant approximation*

In many instances it is sufficient to take the surface charge $\sigma_i(\mathbf{r})$ to be constant within each subdomain as illustrated in Fig. 4.3. This case is dealt with in more detail below:

Equation (2.06) becomes

$$\int_{\Delta s_j} V_0(\mathbf{r}) \, ds = \sum_{i=1}^{M} \int_{\Delta s_j} \int_{\Delta s_i} \sigma_i G(\mathbf{r}|\mathbf{r}') \, ds' \, ds, \tag{2.10}$$

where σ_i is a constant within each subdomain. Write $q_i = \sigma_i \Delta s_i$. Clearly, q_i is the charge on a unit length strip Δs_i wide of one or other of the coaxial conductors. If V_j is the average potential on the jth strip, then from Eq. (2.10),

$$V_j = \sum_{i=1}^{M} \sum_{j=1}^{M} \frac{q_i}{\Delta s_i \, \Delta s_j} \int_{\Delta s_j} \int_{\Delta s_i} G(\mathbf{r}|\mathbf{r}') \, ds' \, ds. \qquad (2.11)$$

This represents a matrix equation of the form

$$\mathbf{V} = \mathbf{PQ}, \qquad (2.12)$$

where in this case the elements of the row vectors $\mathbf{V}$ and $\mathbf{Q}$ have direct significance as voltage and charge. Using Eq. (2.03) the matrix $\mathbf{P}$ is defined by

$$P_{ij} = -\frac{1}{2\pi\varepsilon \, \Delta s_i \, \Delta s_j} \int_{\Delta s_i} \int_{\Delta s_j} \ln \left| \frac{\mathbf{r}_i - \mathbf{r}'_j}{r_0} \right| ds' \, ds. \qquad (2.13)$$

Clearly, the elements defined by Eq. (2.13) can be evaluated exactly for any particular geometry, if necessary by numerical integration. However, in many cases, the subdivision into elements Δs_i will be so fine as to allow the integrand of Eq. (2.13) to be considered constant for each element. In such cases $\mathbf{r}_i$ and $\mathbf{r}'_j$ should be taken as the respective element centroids, $\bar{\mathbf{r}}_i$ and $\bar{\mathbf{r}}_j$. Such an approximation is not valid for the singular case $i = j$. To evaluate P_{ii} it is sufficient to consider a straight-line element aligned with one of the axes:

$$P_{ii} = -\frac{1}{2\pi\varepsilon \, \Delta s_i^2} \int_0^{\Delta s_i} \int_0^{\Delta s_i} \ln \left| \frac{x - x'}{r_0} \right| dx' \, dx. \qquad (2.14)$$

Although Eq. (2.14) has a singular integrand the execution of the double

Fig. 4.3. One-dimensional example of a piecewise constant approximation for $\sigma(r)$. Notice that the width Δs_i may vary between subdomains.

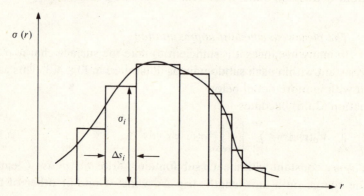

integration is straightforward and gives

$$P_{ii} = -\frac{1}{2\pi\varepsilon}\left[\ln\frac{\Delta s_i}{r_0} - \frac{3}{2}\right].$$ (2.15)

Similar expressions may be worked out for the near-singular cases of adjacent subdomains i and j, although the error incurred in using the simple formula

$$P_{ij} = -\frac{1}{2\pi\varepsilon}\ln\left|\frac{\bar{\mathbf{r}}_i - \bar{\mathbf{r}}_j}{r_0}\right|$$ (2.16)

is never more than a few per cent.

Thus a vector equation

$$\mathbf{Q} = \mathbf{P}\mathbf{V}^{-1}$$ (2.17)

can, in principle, be established to solve the coaxial-capacitor problem. $\mathbf{P}$ is a geometry-dependent matrix whilst $\mathbf{V}$ is a known vector with $V_i = V_A$ or V_B according to whether Δs_i lies on the inner or outer coaxial conductor respectively. Note that the vector $\mathbf{Q}$ is made up from terms $\sigma_i \Delta s_i$, embodying the principle that great advantage is to be had from having a flexible choice of element size.

It can now be seen that the constant r_0, introduced in Eq. (2.01) to express the arbitrary location of reference zero for potential, may in fact always be ignored. For, from Eq. (2.13) it is evident that r_0 appears only as a constant addition $(\ln r_0)/(2\pi\varepsilon)$ to each and every element of $\mathbf{P}$. Thus Eq. (2.12) might be written, in index form, as

$$V_j = \sum_{i=1}^{M} P'_{ij}q_i + \sum_{i=1}^{M}\frac{\ln r_0}{2\pi\varepsilon}q_i.$$ (2.18)

But $\sum_{i=1}^{n}q_i$ is the *total* charge on the system, which must add to zero. Evidently, any finite constant scale-factor r_0 (commonly unity) may be used in expressing the matrix $\mathbf{P}$.

2.3   The rectangular cross-sectioned coaxial line

The principles established above are directly applicable to the rectangular sectioned line examined in Chapter 2, Section 5, for the case where the interconductor space is homogeneous. It was observed in the treatment given in Chapter 2 that the two planes of symmetry could be exploited and, naturally, the same must be the case for the integral equation method here. In Fig. 4.4 it is seen that for every element Δs_i centred about $\mathbf{r}_{1i}$ in the first quadrant and carrying surface charge $q_i = \sigma_i \Delta s_i$ there exist three other elements at points of symmetry $\mathbf{r}_{2i}$, $\mathbf{r}_{3i}$ and $\mathbf{r}_{4i}$ in the remaining quadrants, each carrying exactly the same charge. Consequently, the number of unknowns q_i, $i = 1, 2, \ldots, N$, need only correspond to division of the boundary in the first quadrant. Clearly, in setting up an equation such as Eq. (2.11) (for the piecewise-constant-charge approximation), such a reduced number of unknowns may be exploited provided the Green's function employed in Eq. (2.13) is now taken as

$$G(\mathbf{r}|\mathbf{r}') = -\{\ln|\mathbf{r}_{1i} - \mathbf{r}'_j| + \ln|\mathbf{r}_{2i} - \mathbf{r}'_j|$$
$$+ \ln|\mathbf{r}_{3i} - \mathbf{r}'_j| + \ln|\mathbf{r}_{4i} - \mathbf{r}'_j|\}/2\pi\varepsilon. \qquad (2.19)$$

3.   Green's function for a dielectric slab

One of the drawbacks encountered with integral operator methods is the fact that the Green's functions involved depend very much upon the particular problem being tackled. So far only the very simple situation of a uniform medium has been dealt with. Now a dielectric slab is considered and it is shown how a Green's function may be found for such a geometry by the method of partial images.

3.1   *Dielectric half-space*

The simple case of a half-space, $x < 0$, filled with a homogeneous dielectric of permittivity ε_1 is considered. Two-dimensional symmetry is assumed and a line charge q coulombs per metre is supposed lying at distance a from the surface $x = 0$. This charge results in electric flux radiating uniformly from its length. Consider a typical tube of flux ψ as shown in Fig. 4.5. At the dielectric surface, some fraction of the flux $K\psi$ will fail to penetrate, whilst the remainder $(1 - K)\psi$ must continue

Fig. 4.4. Rectangular sectioned coaxial line.

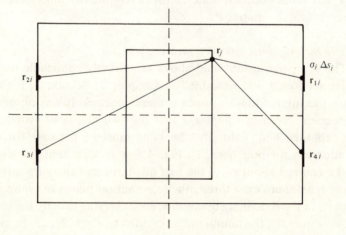

into the dielectric material. The fraction K and the direction taken by the flux lines which fail to penetrate the interface may be found from the conditions of continuity for electrostatic fields which were discussed in Chapter 2, Section 1.3. These require that the normal component of flux density and the tangential component of electric field should be continuous across the interface. From Fig. 4.5 it is seen that the normal flux density requirement gives

$$(1 - K)\psi \sin \alpha_1 = \psi \sin \alpha_1 - K\psi \sin \alpha_2. \tag{3.01}$$

It follows that $\alpha_1 = \alpha_2$ always; that is, the angles subtended by the incident and returned flux must be equal. On the other hand, continuity of the tangential electric field component is possible only if

$$\frac{1}{\varepsilon_1}(1 - K)\psi \cos \alpha_1 = \frac{1}{\varepsilon_0}\psi \cos \alpha_1 + \frac{1}{\varepsilon_0}K\psi \cos \alpha_2, \tag{3.02}$$

so that K must have the value

$$K = \frac{\varepsilon_0 - \varepsilon_1}{\varepsilon_0 + \varepsilon_1}. \tag{3.03}$$

The analysis above shows that the geometrical relationships governing the behaviour of flux lines near a dielectric interface are analogous to those of optics and that the image coefficient K plays a role corresponding to the optical reflection coefficient. The equality of angles found leads to an apparent existence of images as shown in Fig. 4.5. Flux lines on the right-hand side of the interface appear to be due to two different line sources, the original source q and an image source Kq located in the dielectric region at a distance a behind the interface. Now considering the region $x < 0$, any measurement performed in this dielectric region would indicate a single source of strength $(1 - K)q$ located in the right

Fig. 4.5. Typical flux line associated with a line charge near a semi-infinite dialectric half-space.

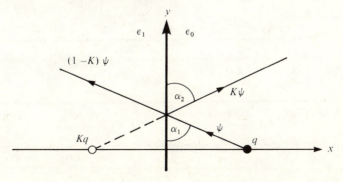

half-space. Thus, for the simple half-space dielectric slab, the potential of the line source of Fig. 4.5 is as follows:

$x > 0$:

$$V(x, y) = -\frac{q}{4\pi\varepsilon}\left[\ln\{(x-a)^2 + y^2\} + K\ln\{(x+a)^2 + y^2\}\right], \quad (3.04)$$

$x < 0$:

$$V(x, y) = -\frac{q}{4\pi\varepsilon}(1-K)\ln\{(x-a)^2 + y^2\}. \quad (3.05)$$

3.2   *The dielectric slab*

The method used here to describe image formation in the single interface of the dielectric half-space is immediately applicable to the

Fig. 4.6. (*a*) **Flux lines due to a line charge near a dialectric slab.**
(*b*) **Image representation valid in the region containing the charge.**
(*c*) **Image representation valid in the central region. (*d*) Image representation valid in the left region.**

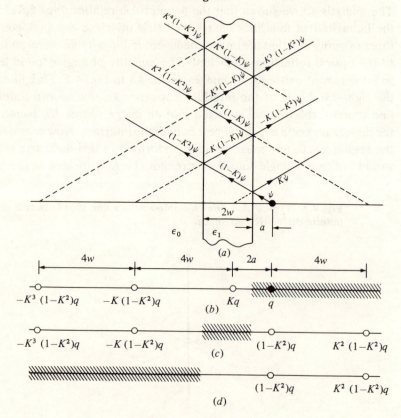

slab of finite thickness. Figure 4.6 shows how multiple images arise, with a different imaging for each of the three regions corresponding to the slab. The strengths and positions of the image charges may readily be checked by the reader using the principles which were established for imaging in the infinite half-space case (Fig. 4.5). Having obtained the image pattern for a dielectric slab, Green's functions, valid in each of the three regions separated by the slab faces, are easily written down from the resulting infinite-series potential function

$$V(\mathbf{r}) = - \sum_{i=1}^{\infty} \frac{1}{2\pi\varepsilon} q_i \ln |\mathbf{r} - \mathbf{r}_i|, \tag{3.06}$$

where q_i is the charge associated with an ith image (or the original-line charge where applicable) located at position $\mathbf{r}_i$. Note that the appropriate value of ε, either ε_1 or ε_0, corresponds to the location of the observation point $\mathbf{r}$, and *not* to the image point $\mathbf{r}_i$.

3.3   *Microstrip transmission-line analysis*

The calculation of parameters associated with microstrip line can be achieved by the methods outlined here. It can be seen immediately how the basic microstrip problem, that of calculating the capacitance between dielectric-separated ribbons as illustrated in Fig. 4.7, might be tackled. The conducting ribbons are discretised as were the surfaces of the coaxial line with uniform dielectric. The matrix relationship, Eq. (2.12), is set up as before, this time with the Green's function in Eq. (2.11) corresponding to the infinite series of line charges Eq. (3.06). The actual-line capacitance may be determined as the ratio of charge to voltage by summing the individual charges calculated from solution of Eq. (2.12) for one or other of the electrodes separately.

Fig. 4.7. Microstrip transmission line. (*a*) Ribbon conductor and dielectric sheet with conducting backing. (*b*) Electrically equivalent 2-ribbon line.

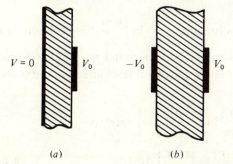

(*a*)   (*b*)

4.   Variational expressions for integral operators

In previous chapters dealing with differential operators, the dominant theme behind the analysis developed has always been the variational principle. Here, however, in the first instance the corresponding integral operators have been exploited directly. The direct method yields results in rather a simple fashion when zero-order finite elements are employed, for instance in the example of Section 2.2, with constant charge density on individual boundary elements. The employment of higher-order approximations was shown in Section 2.1 to be a more complex matter. In this Section it is shown how the integral equation used for the electrostatics problem can quite conveniently be cast into variational form. Continuing to consider the uniform coaxial-line problem of Section 2, it is readily found that the functional

$$F(\sigma) = \int_S V_0(\mathbf{r})\sigma(\mathbf{r})\,ds$$

$$+ \tfrac{1}{2}\int_S\int_S \frac{\sigma(\mathbf{r})\sigma(\mathbf{r}')}{2\pi\varepsilon}\ln\left|\frac{\mathbf{r}-\mathbf{r}'}{r_0}\right|\,ds\,ds' \tag{4.01}$$

is stationary to perturbations about the true charge distribution. This may be demonstrated by examining Eq. (4.01) when a specific charge distribution σ is replaced by $\sigma + \delta\sigma$, $\delta\sigma$ being an arbitrary perturbation of σ and function of the position vector $\mathbf{r}$ on the contour S. Writing $F(\sigma + \delta\sigma) = F(\sigma) + \delta F$ it follows immediately that

$$\delta F = \int_S V_0(\mathbf{r})\delta\sigma(\mathbf{r})\,ds$$

$$+ \tfrac{1}{2}\int_S\int_S \frac{\sigma(\mathbf{r})\delta\sigma(\mathbf{r}')}{2\pi\varepsilon}\ln\left|\frac{\mathbf{r}-\mathbf{r}'}{r_0}\right|\,ds\,ds'$$

$$+ \tfrac{1}{2}\int_S\int_S \frac{\sigma(\mathbf{r}')\delta\sigma(\mathbf{r})}{2\pi\varepsilon}\ln\left|\frac{\mathbf{r}-\mathbf{r}'}{r_0}\right|\,ds\,ds'$$

$$+ \tfrac{1}{2}\int\int \frac{\delta\sigma(\mathbf{r})\delta\sigma(\mathbf{r}')}{2\pi\varepsilon}\ln\left|\frac{\mathbf{r}-\mathbf{r}'}{r_0}\right|\,ds\,ds'. \tag{4.02}$$

It is observed that the Green's function $\dfrac{1}{2\pi\varepsilon}\ln\left|\dfrac{\mathbf{r}-\mathbf{r}'}{r_0}\right|$ is perfectly symmetrical in the position vectors $\mathbf{r}$ and $\mathbf{r}'$ appearing in Eq. (4.02) as dummy variables of integration. Thus $\mathbf{r}$ and $\mathbf{r}'$ may be interchanged in the second integral of Eq. (4.02), giving

$$\delta F = \int_S \left[V_0(\mathbf{r}) + \int_S \frac{\sigma(\mathbf{r}')}{2\pi\varepsilon}\ln\left|\frac{\mathbf{r}-\mathbf{r}'}{r_0}\right|\,ds' \right]\delta\sigma(\mathbf{r})\,ds$$

$$+ \tfrac{1}{2}\int_S\int_S \frac{\delta\sigma(\mathbf{r})\delta\sigma(\mathbf{r}')}{2\pi\varepsilon}\ln\left|\frac{\mathbf{r}-\mathbf{r}'}{r_0}\right|\,ds\,ds'. \tag{4.03}$$

In Eq. (4.03) the term linear in $\delta\sigma$, the *first variation* of F, vanishes identically if $\sigma(\mathbf{r})$ itself is a solution of the basic integral equation, Eq. (2.02), which has been established for the coaxial-line system. Hence, the functional F expressed by Eq. (4.01) is indeed stationary for perturbations about the true solution σ.

The integral operator expression (4.01) may be used in finite element schemes in a similar fashion to the functionals already identified with the electromagnetics partial differential operators. It is important to note that the finite element here is one-dimensional, a segment of the curve S, whereas the same problem treated by differential operator methods (Chapter 2, Section 5) requires the employment of two-dimensional elements. Taking the approach established for the differential operator functionals, it should be possible to define finite element systems having the variable (σ in this case) follow some prescribed law within each element, say linear, quadratic or even higher order, rather than the constancy assumed in Section 2.2. Nodal values of the variable may be fitted to the prescribed law and the functional F set up in terms of these nodal quantities. The particular values of these quantities making F stationary may be taken as giving the best representation possible of the trial function system to the true solution.

5.   Integral equations in magnetostatics

The advantages to be gained by using integral equation formulation of field problems are particularly important in magnetostatics analyses. There is much interest in accurately solving magnetic field problems associated with the rather detailed geometries of electric machines. A number of programs are in existence which exploit the reduction in dimensionality and the simplification with respect to open boundary conditions which may be obtained using integral methods in comparison with the differential operator approach.

5.1   *Magnetisation integral equation*

The approach considered here is worked in terms of the *magnetisation vector*

$$\mathbf{M} = \frac{\mathbf{B}}{\mu_0} - \mathbf{H}. \tag{5.01}$$

The vector $\mathbf{M}$ is a measure of the internal magnetism present in a material, either as permanent magnetism or induced through the presence of current sources. It may be noted that $\mu = \mu_0$, $\mathbf{B} = \mu_0\mathbf{H}$ correspond to a nonmagnetic medium and Eq. (5.01) confirms that $\mathbf{M}$ vanishes under these circumstances. Equation (5.01) may be written

$$\mathbf{M} = \chi \mathbf{H}, \tag{5.02}$$

from which it follows that the *susceptibility* χ is given by

$$\chi = \mu/\mu_0 - 1. \tag{5.03}$$

This parameter may be regarded as a known material property. In many cases χ will be dependent upon $\mathbf{H}$. The magnetic field intensity in Eq. (5.01) can be separated into two parts

$$\mathbf{H} = \mathbf{H_S} + \mathbf{H_M}, \tag{5.04}$$

the first part, $\mathbf{H_S}$, being directly the result of the current sources present whereas $\mathbf{H_M}$ is regarded as being due to magnetisation induced in the material. If it is assumed that $\mathbf{H_S}$ is entirely independent of the material properties then this vector must be determined by the relationship applying to distributed currents in free space,

$$\mathbf{H_S}(\mathbf{r}) = \frac{1}{4\pi} \int_{\Omega_J} \frac{\mathbf{J}(\mathbf{r'}) \times (\mathbf{r} - \mathbf{r'})}{|r - r'|^3} \, d\Omega'. \tag{5.05}$$

Equation (5.05) is effectively a statement of the well-known Biot–Savart law (see Chapter 2, Section 2). The vectors $\mathbf{r}$ and $\mathbf{r'}$ are, respectively, field and source point vectors OP and OQ referred to a common origin O, as depicted in Fig. 4.1. The integration is over the region of space Ω_J which contains all the current sources and the operation is performed with respect to the variable $\mathbf{r'} = (x', y', z')$.

The Biot–Savart equation Eq. (5.05) is in fact a solution of the equation

$$\nabla \times \mathbf{H_S} = \mathbf{J}. \tag{5.06}$$

The reader is referred to standard works on electromagnetism, for example Stratton (1941), p. 225 *et seq.*, for detailed proof of this proposition. Since the Maxwell equation,

$$\nabla \times \mathbf{H} = \mathbf{J}, \tag{5.07}$$

applying to time-invariant situations must hold, it follows that $\mathbf{H_M}$ is *irrotational*, $\nabla \times \mathbf{H_M} = 0$. By the rules of vector calculus this property is satisfied identically if $\mathbf{H_M}$ is chosen as the gradient of a scalar potential function P,

$$\mathbf{H_M} = -\nabla P. \tag{5.08}$$

The quantity P is described as the *reduced scalar potential* since it represents only part of the magnetic field. It becomes the same as the scalar magnetic potential defined in Chapter 2, Section 2.2 for regions where current is absent. From Eq. (5.01) it follows that

$$\mathbf{H_M} + \mathbf{H_S} = \frac{\mathbf{B}}{\mu_0} - \mathbf{M}. \tag{5.09}$$

The divergence of $\mathbf{H_S}$ vanishes since this vector represents a magnetic field in free space. Thus, taking the divergence of Eq. (5.09) reveals that

$$\nabla \cdot \mathbf{H_M} = -\nabla \cdot \mathbf{M}. \tag{5.10}$$

Then from Eqs. (5.08) and (5.10) it is seen that the governing partial differential equation for the reduced scalar potential is the Poisson equation

$$\nabla^2 P = \nabla \cdot \mathbf{M}. \tag{5.11}$$

Equation (5.11) may be compared with Poisson's equation as it appears in electrostatics for a uniform material:

$$\nabla^2 V = -\rho/\varepsilon. \tag{5.12}$$

The electrostatics solution of Eq. (5.12) has already been discussed and, for a volume Ω bounded by a surface S, is

$$V(\mathbf{r}) = \int_\Omega \frac{\rho(\mathbf{r}')\, d\Omega'}{4\pi\varepsilon |\mathbf{r} - \mathbf{r}'|} + \int_S \frac{\rho_S(\mathbf{r}')\, dS'}{4\pi\varepsilon |\mathbf{r} - \mathbf{r}'|}. \tag{5.13}$$

It is noted that the volume and surface charges of Eq. (5.13) may be represented in terms of a flux density $\mathbf{D}$ by $\rho = \nabla \cdot \mathbf{D}$, $\rho_S\, dS = -\mathbf{D} \cdot d\mathbf{S}$.

Evidently, Eq. (5.13) can be rewritten as

$$V(\mathbf{r}) = \int_\Omega \frac{\nabla' \cdot \mathbf{D}(\mathbf{r}')\, d\Omega'}{4\pi\varepsilon |\mathbf{r} - \mathbf{r}'|} - \int_S \frac{\mathbf{D}(\mathbf{r}') \cdot d\mathbf{S}'}{4\pi\varepsilon |\mathbf{r} - \mathbf{r}'|}. \tag{5.14}$$

It is necessary to distinguish very carefully between the variable of integration $\mathbf{r}'$ and the point $\mathbf{r}$ at which the potential V is evaluated. The symbol ∇' denotes the vector operator $(\partial/\partial x', \partial/\partial y', \partial/\partial z')$ acting with respect to the primed coordinate system $\mathbf{r}' = (x', y', z')$. Thus, observing that $\nabla \cdot \mathbf{M}$ corresponds to $-\rho/\varepsilon = -\nabla \cdot \mathbf{D}/\varepsilon$ and $\mathbf{M}$ to $-\mathbf{D}/\varepsilon$, it is clear that the solution of Eq. (5.11) for the reduced scalar magnetic potential is

$$P(\mathbf{r}) = -\frac{1}{4\pi} \int_{\Omega_M} \frac{\nabla' \cdot \mathbf{M}(\mathbf{r}')\, d\Omega'}{|\mathbf{r} - \mathbf{r}'|} + \frac{1}{4\pi} \int_{S_M} \frac{\mathbf{M}(\mathbf{r}') \cdot d\mathbf{S}'}{|\mathbf{r} - \mathbf{r}'|}, \tag{5.15}$$

where all the magnetic material is contained within the volume Ω_M enclosed by the surface S_M.

The divergence theorem of vector calculus (Riley (1974), p. 100) may be used to transform the second integral of Eq. (5.15) so that

$$P(\mathbf{r}) = -\frac{1}{4\pi} \int_{\Omega_M} \frac{\nabla' \cdot \mathbf{M}(\mathbf{r}')\, d\Omega'}{|\mathbf{r} - \mathbf{r}'|} + \frac{1}{4\pi} \int_{\Omega_M} \nabla' \cdot \left[\frac{\mathbf{M}(\mathbf{r}')}{|\mathbf{r} - \mathbf{r}'|} \right] d\Omega'. \tag{5.16}$$

The second integral in Eq. (5.16) above may be expanded using the vector calculus rule $\nabla(\psi\mathbf{F}) = \psi\nabla \cdot \mathbf{F} + (\nabla\psi) \cdot \mathbf{F}$ so that, observing a cancellation of terms arising after the expansion, there results

$$P(\mathbf{r}) = \frac{1}{4\pi} \int_{\Omega_M} \mathbf{M}(\mathbf{r}') \cdot \nabla' \left(\frac{1}{|\mathbf{r} - \mathbf{r}'|} \right) d\Omega'. \tag{5.17}$$

Equation (5.17) is written more explicitly

$$P(\mathbf{r}) = \frac{1}{4\pi} \int_{\Omega_M} \frac{\mathbf{M}(\mathbf{r}') \cdot (\mathbf{r} - \mathbf{r}')}{|\mathbf{r} - \mathbf{r}'|^3} \, d\Omega'. \tag{5.18}$$

Since $\mathbf{H}_M$ derives from the gradient of P, Eq. (5.18) yields

$$\mathbf{H}_M(\mathbf{r}) = -\frac{1}{4\pi} \nabla \int_{\Omega_M} \frac{\mathbf{M}(\mathbf{r}') \cdot (\mathbf{r} - \mathbf{r}')}{|\mathbf{r} - \mathbf{r}'|^3} \, d\Omega', \tag{5.19}$$

the gradient operator ∇ here of course being with respect to the unprimed coordinates $\mathbf{r} = (x, y, z)$. Thence, Eq. (5.02) can be written, with the aid of Eqs. (5.04), (5.05) and (5.19),

$$M(\mathbf{r}) = \frac{\chi(\mathbf{r})}{4\pi} \left[\int_{\Omega_J} \frac{\mathbf{J}(\mathbf{r}') \times (\mathbf{r} - \mathbf{r}')}{|\mathbf{r} - \mathbf{r}'|^3} \, d\Omega' \right.$$
$$\left. - \nabla \int_{\Omega_M} \frac{\mathbf{M}(\mathbf{r}') \cdot (\mathbf{r} - \mathbf{r}')}{|\mathbf{r} - \mathbf{r}'|^3} \, d\Omega' \right]. \tag{5.20}$$

This is an integral equation which determines the unknown vector $\mathbf{M}$ given χ and $\mathbf{J}$ everywhere.

5.2   *Application of the magnetisation integral equation*

The problem of determining the magnetic field of a current-carrying conductor close to and parallel to a long bar of ferromagnetic material is considered. In this two-dimensional case there is no variation in the axial (say z) direction. Equation (5.18) can be integrated at once with respect to z' to give the scalar magnetic reduced potential as

$$P(x, y) = \frac{1}{2\pi} \iint \frac{\mathbf{M}(\mathbf{r}') \cdot (\mathbf{r} - \mathbf{r}')}{|\mathbf{r} - \mathbf{r}'|^2} \, dx' \, dy', \tag{5.21}$$

where now the symbols $\mathbf{r}$ and $\mathbf{r}'$ imply the two-dimensional radius vectors of field and source points, (x, y) and (x', y') respectively. Applying the gradient operation, Eq. (5.08), to the two-dimensional expression (5.21) above then gives

$$\mathbf{H}_M = -\frac{1}{2\pi} \iint \left\{ \frac{\mathbf{M}(\mathbf{r}')}{|\mathbf{r} - \mathbf{r}'|^2} - [2\mathbf{M}(\mathbf{r}') \cdot (\mathbf{r} - \mathbf{r}')] \frac{(\mathbf{r} - \mathbf{r}')}{|\mathbf{r} - \mathbf{r}'|^4} \right\} \, dx' \, dy'. \tag{5.22}$$

A particular geometry, that of a pair of wires carrying current parallel to a rectangular bar of iron, is chosen to illustrate the problem and shown in Fig. 4.8. The source field $\mathbf{H}_S$ at the point (x_k, y_k) can be written down immediately

$$H_{Sk}^{(x)} = \frac{I}{2\pi} \left[\frac{y_S - y_k}{x_k^2 + (y_k - y_S)^2} + \frac{y_S + y_k}{x_k^2 + (y_k + y_S)^2} \right], \tag{5.23}$$

$$H_{Sk}^{(y)} = \frac{I}{2\pi} \left[\frac{x_k}{x_k^2 + (y_k - y_S)^2} - \frac{x_k}{x_k^2 + (y_k + y_S)^2} \right]. \tag{5.24}$$

In a general case without the symmetry exhibited here, suppose that the iron cross-section is divided into N elements. Let an arbitrary point within the lth element be $\mathbf{r}_l'$, $l = 1, 2, \ldots, N$, and suppose as an approximation that the magnetisation is constant within each element. Then also as an approximation, which becomes increasingly more accurate as the number of elements increases, it is clear that the magnetisation at a point $\mathbf{r}_k$, $k = 1, 2, \ldots, N$, can be written as the sum of contributions from each of the elements $1 = 1, 2, \ldots, N$ (including one from the kth element, in which it is assumed that the field point $\mathbf{r}_k$ lies). Using Eq. (5.22) this sum is

$$\mathbf{H}_{Mk} = -\frac{1}{2\pi} \sum_{l=1}^{N} \iint \left[\frac{\mathbf{M}_l}{|\mathbf{r}_k - \mathbf{r}_l'|^2} - \frac{2\mathbf{M}_l \cdot (\mathbf{r}_k - \mathbf{r}_k')}{|\mathbf{r}_k - \mathbf{r}_l'|^4} (\mathbf{r}_k - \mathbf{r}_l') \right] dx_l' \, dy_l'. \tag{5.25}$$

Since the magnetisation vector $\mathbf{M}_l$ is assumed to be constant within each element, the integrations of Eq. (5.25), although algebraically complicated, can be performed (Newman, Trowbridge & Turner, 1972). The result is purely dependent upon the element geometry, which need not necessarily be rectangular as chosen here. The integration over each individual element must be performed relative to some reference point $\mathbf{r}_l$. If the reference points for integration $\mathbf{r}_l$, $l = 1, 2, \ldots, N$, are chosen to be the same as the field points $\mathbf{r}_k$, it is clear that by this assumption of *collocation*, an equation,

$$\mathbf{H}_{Mk} = \sum_{l=1}^{N} \mathbf{C}_{kl} \mathbf{M}_l \quad k = 1, 2, \ldots, N, \tag{5.26}$$

Fig. 4.8. Current-carrying wires parallel to a bar of iron.

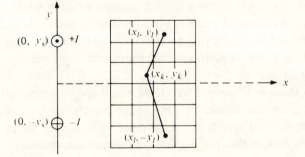

can be written. The symbol $\mathbf{C}_{kl}$ is intended to represent a *second-order tensor*

$$\mathbf{C}_{kl} = \begin{bmatrix} C_{kl}^{xx} & C_{kl}^{xy} \\ C_{kl}^{yx} & C_{kl}^{yy} \end{bmatrix}, \tag{5.27}$$

which in Eq. (5.26) operates upon a vector to produce yet another vector, whilst

$$\mathbf{C}_{kl}\mathbf{M}_l = [C_{kl}^{xx}M_i^x + C_{kl}^{xy}M_i^y, \ C_{kl}^{yx}M_i^x + C_{kl}^{yy}M_i^y]. \tag{5.28}$$

The collocation point may be chosen anywhere within its element. However, in order to make the approximation as efficient as possible for a given size of element, it is intuitively apparent that the element centroid is an appropriate choice.

Thus the problem of determining the magnetisation has been discretised, for now it may be written, from Eqs. (5.02) and (5.04),

$$\mathbf{M}_k = \chi(\mathbf{r}_k)\left[\mathbf{H}_{Sk} + \sum_{l=1}^{N} \mathbf{C}_{kl}\mathbf{M}_l\right] \tag{5.29}$$

or

$$\sum_{l=1}^{N}[\mathbf{C}_{kl} + \delta_{kl}/\chi(\mathbf{r}_k)]\mathbf{M}_l = \mathbf{H}_{Sk}, \tag{5.30}$$

where the *Kronecker delta*, $\delta_{kl} = 1$, $k = l$ or $\delta_{kl} = 0$, $k \neq l$, is introduced. Equation (5.30) is of matrix form and can be inverted to yield $\mathbf{M}_l$.

The exploitation of symmetry such as displayed in the example of Fig. 4.4 is now considered. Here, by symmetry, the magnetisation at points (x_l, y_l) and $(x_l, -y_l)$ is the same. Thus the number of unknown $\mathbf{M}_l$ needs only to correspond to the elements in the region $y > 0$. Equation (5.25), which gave the coefficients $\mathbf{C}_{kl}$, now is modified to include additional terms obtained replacing $\mathbf{r}'_l = (x'_l, y'_l)$ by $(x'_l, -y'_l)$. The summation proceeds to obtain $\mathbf{H}_{Ml}$ counting $l = 1, 2, \ldots, N$ to correspond with the elements in the region $y_l > 0$ only.

If it is wished to take into account nonlinear material properties, Eq. (5.30) can be solved iteratively starting from a guessed solution which, given B–H-curves for the material, enables $\chi(\mathbf{r}_k)$ to be estimated initially.

The method described here applying to a two-dimensional problem is readily extended to three dimensions and forms the basis for the GFUN program described by Newman *et al.* (1972). It is a viable alternative to the methods of Chapter 5 which are based upon variational solution of partial differential equations. One of the principal merits of the integral equation method here is that the division of space into finite elements is necessary only in the iron portions of the problem-space.

6.   Finite elements in antenna theory

The design and analysis of radio antennas is a well-established topic in electrical engineering and is based upon electromagnetic-wave theory. Antenna design and analysis was given substantial new impetus with the advent of digital electronic computing, enabling many hitherto intractable problems to be tackled with comparative ease. The most important numerical technique in the solution of the antenna integral equations is collocation, also known as the *method of moments*. Since this is essentially a finite element technique, brief mention of the topic is made here. The reader seriously interested in antennas should of course consult one of the specialist works on the subject, for instance the text by Balanis (1982).

The problem which an antenna designer invariably has to face is the calculation of the distributed currents in some antenna array. Each member of the array is either driven directly or is excited through being in the electromagnetic field of other elements. Once the current distribution has been fully determined there is then no fundamental problem in obtaining the radiation pattern of a transmitting array. It may, for instance, be calculated via Eq. (2.09) of Chapter 2, even though the arithmetic involved in integrating the expression for vector potential **A** may be heavy. Alternatively, knowledge of the currents everywhere establishes the circuit behaviour of a receiving antenna.

6.1   *Pocklington's equation*

This integral equation is one of the most useful in the numerical analysis of linear antennas and, remarkably, dates from a time when electromagnetics theory was still in its infancy.[1] A straight cylindrical conducting wire with small cross-section, both compared with its length and with the free-space electromagnetic wavelength, is considered. The very reasonable assumption that the current is entirely axially directed with azimuthal symmetry is made. End effects are ignored and, corresponding to perfect conductivity, all currents are assumed confined to the cylindrical surface. In modern notation, Pocklington's result is obtained by noting that the vector potential **A** due to axial current on

---

[1] H. C. Pocklington, 'Electrical oscillations in wires', *Cambridge Phil. Soc. Proc.*, vol. 9 (1897), pp. 324–32. It is interesting to note that in this same volume J. J. Thomson publishes a paper on cathode rays, from which it is clear that he was then just a little short of deducing the existence of the electron. Immediately following Pocklington's paper, C. T. R. Wilson writes on experiments in his cloud chamber, detecting rays emanating from uranium.

the cylinder must itself be entirely z-directed. Using the retarded potential solution of Maxwell's equations indicated in Chapter 2 (Eq. (2.09) there), and referring to Fig. 4.9, this sole component is given by an integral over the volume of the cylinder

$$A_z(x, y, z) = \mu_0 \int \frac{J_z(x', y', z') \, e^{-jkR}}{4\pi R} \, d\Omega', \tag{6.01}$$

where the complex phasor convention at frequency ω is followed and $k = \omega(\mu_0 \varepsilon_0)^{1/2}$. In Eq. (6.01) A_z is expressed in the first place as if varying in the x–y-plane and the volume integral is taken with respect to dx', dy' as well as dz'. However, since the current is uniformly distributed upon the cylinder surface at radius a, Eq. (6.01) gives to good approximation a value referring to the surface of the cylinder

$$A_z(z) = \mu_0 \int_{-l/2}^{l/2} \frac{I(z') \, e^{-jk\bar{R}}}{4\pi \bar{R}} \, dz', \tag{6.02}$$

where a mean value of R over the x', y' integration

$$\bar{R} = [a^2 + (z' - z)^2]^{1/2} \tag{6.03}$$

Fig. 4.9. Cylindrical antenna.

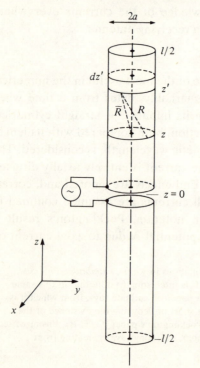

is used in Eq. (6.02). Equation (2.03) of Chapter 2 indicates that here the z-directed electric field at the cylinder surface due to the current distribution described is

$$E_z^s = -\frac{\partial V}{\partial z} - j\omega A_z, \tag{6.04}$$

where V is the scalar electric potential. But assuming the Lorentz gauge (Chapter 2, Eq. (2.05)),

$$\frac{\partial A_z}{\partial z} = -\mu_0 \varepsilon_0 j\omega V. \tag{6.05}$$

Thus

$$j\omega\varepsilon_0 E_z^s = \frac{1}{\mu_0}\frac{\partial^2 A_z}{\partial z^2} + \omega^2 \varepsilon_0 A_z. \tag{6.06}$$

Eliminating A_z between Eqs. (6.02) and (6.06) gives Pocklington's integral equation for the current distribution $I(z')$

$$j\omega\varepsilon_0 E_z^s = \int_{l/2}^{l/2} \left[I(z')\left(\frac{\partial^2}{\partial z^2} + k^2\right)\frac{e^{-jk\bar{R}}}{4\pi\bar{R}} \right] dz'. \tag{6.07}$$

6.2   Finite element solution

If the cylinder concerned is isolated but driven, say at $z = 0$, then E_z^s, being the tangential field at a perfect conductor surface, vanishes everywhere except over the cylinder surface (produced) of the driven gap. The simplest numerical solution is obtained by dividing the antenna half-length $l/2$ into, say, M finite elements over each of which $I(z')$ is assumed to be constant, say $I_1, I_2, \ldots, I_n, \ldots, I_M$. Substituting these constant values into Eq. (6.07) and performing the integration over z', M values of z may also be chosen, one within each individual finite element, at which E_z is known. The value of E_z will, in this example, be zero except for the driven region. Clearly, the matrix equation

$$\mathbf{V} = \mathbf{ZI} \tag{6.08}$$

results, which can be solved for the unknown currents $\mathbf{I} = [I_1, I_2, \ldots, I_M]$.

This procedure is exactly as was described in more detail for the electrostatics case and corresponds to the zero-order Galerkin solution. Just as in that case, higher-order approximations may be assumed and the full Galerkin procedure executed. Once the current distribution $I(z')$ has been determined in whatever approximation, practical parameters such as the driven impedance and the radiation pattern are then readily worked out. The cylinder under discussion might represent part of a multiple array, some elements being parasitic and, of course, at least one being specifically driven. Then E_z^s expresses just the field due to current in the element itself. To it must be added the impressed field E_z^i arising from current in neighbouring elements and, except at the driving points, the sum must vanish:

$$E_z^i + E_z^s = 0. \tag{6.09}$$

E_z^s in Eq. (6.05) is thus determinable by means of this relation. The matrix system of equations will be correspondingly enlarged but, nevertheless, still solvable. The theory given here applies to individual elements which are straight and uniform. However, it can be seen that, at the expense of some complexity in programming, nonuniform and curved arrays can be synthesised.

7.   Readings

The book by Harrington (1968) is an early reference dealing with some fundamentals of field computation by what would now be described as finite element techniques. In particular, solution of the integral equations for electrostatics problems is discussed at some length. However, the treatment of the electrostatics problems given here is based upon papers by Silvester (1968) and Benedek & Silvester (1972). Finite element methods of high order are applied to wire-antenna problems by Silvester & Chan (1972) and to reflector antennas by Hassan & Silvester (1977). The research monograph edited by Chari & Silvester (1980) contains a chapter by C. W. Trowbridge, 'Applications of integral equation methods to magnetostatics and eddy-current problems', which is relevant to Section 5 here. In the same monograph B. H. McDonald and A. Wexler give some rather advanced theory concerning the treatment of mixed partial differential and integral equation formulations. The integral equations of antenna theory are discussed by Balanis (1982) in his comprehensive work covering antenna theory and practice.

References

Balanis, C. A. (1982). *Antenna Theory, Analysis and Design*. New York: Harper and Row.

Benedek, P. & Silvester, P. P. (1972). 'Capacitance of parallel rectangular plates separated by a dielectric sheet', *Institute of Electrical and Electronics Engineers Transactions on Microwave Theory and Techniques*, **MTT–20**, 504–10.

Chari, M. V. K. & Silvester, P. P. (1980). *Finite Elements in Electric and Magnetic Field Problems*. Chichester: John Wiley.

Harrington, R. F. (1968). *Field Computation by Moment Methods*. New York: Macmillan.

Hassan, M. A. & Silvester, P. (1977). 'Radiation and scattering by wire antenna structures near a rectangular plate reflector', *Institution of Electrical Engineers Proceedings*, **124**, 429–35.

Newman, M. J., Trowbridge, C. W. & Turner, L. R. (1972). 'GFUN: an interactive program as an aid to magnet design', *Proceedings of the 4th International Conference on Magnet Technology*, Brookhaven, pp. 617–26.

Riley, K. F. (1974). *Mathematical Methods for the Physical Sciences*. Cambridge University Press.

Silvester, P. (1968). 'TEM properties of microstrip transmission lines', *Institution of Electrical Engineers Proceedings*, **115**, 43–8.

Silvester, P. & Chan, K. K. 1972. 'Bubnov–Galerkin solutions to wire-antenna problems', *Institution of Electrical Engineers Proceedings*, **119**, 1095–9.

Stratton, J. A. (1941). *Electromagnetic Theory*. New York: McGraw-Hill.

5

Differential operators in ferromagnetic materials

1. Functionals for the magnetic field

Areas of electromagnetics analysis in which finite elements have to date found extensive use include problems of nonlinear magnetics (with application to electric machines, transformers, and other power devices). The nonlinearities encountered in such problems are usually single-valued and monotonic or at least can be assumed to be such. Comparatively simple methods can therefore be made computationally efficient.

As set out in Chapter 2, magnetic field problems are commonly described either in terms of a scalar potential P, so defined that the magnetic field H is its gradient

$$\mathbf{H} = -\nabla P, \tag{1.01}$$

or in terms of a vector potential $\mathbf{A}$, whose curl gives the magnetic flux density $\mathbf{B}$:

$$\mathbf{B} = \nabla \times \mathbf{A}. \tag{1.02}$$

The scalar potential can be defined readily only in regions where $\mathbf{H}$ is irrotational, i.e., in current-free regions. There, P is governed by the nonlinear differential equation

$$\nabla \cdot (\mu \nabla P) = 0, \tag{1.03}$$

which is nonlinear because the material permeability is a function of the magnetic field itself. Correspondingly, the vector potential is governed by

$$\nabla \times (\nu \nabla \times \mathbf{A}) = \mathbf{J}, \tag{1.04}$$

which may be regarded as more general because the problem region may include nonzero current densities $\mathbf{J}$. These may of course represent imposed current densities as well as current densities which result from time variations of the fields themselves (eddy currents). The material reluctivity ν (the inverse of material permeability) is again field dependent, hence Eq. (1.04) is nonlinear.

Equations (1.03)–(1.04) are valid for general three-dimensional fields. Although there is considerable practical interest in three-dimensional problems, a great majority of practical cases tends to be handled by two-dimensional approximation – not only because computing costs are high for three-dimensional cases, but also because the representation of results and of geometric input data give rise to serious problems. For example, electric machines are often analysed on the assumption that they have infinite axial length. In such cases, Eq. (1.03) still applies, subject to the understanding that the divergence and gradient operators refer to two dimensions. In Eq. (1.04), the vectors may be assumed to be entirely z-directed, so that (1.04) reduces to the two-dimensional form

$$\frac{\partial}{\partial x}\left(\nu\frac{\partial A}{\partial x}\right)+\frac{\partial}{\partial y}\left(\nu\frac{\partial A}{\partial y}\right)=-J, \tag{1.05}$$

which is formally identical to a two-dimensional inhomogeneous version of (1.03), except for the obvious changes of variable. Indeed, (1.05) has precisely the same general appearance as linear problems in Poisson's equation; the difference lies only in the dependence of material properties on the field.

It is important to note that the permeabilities of most common materials may be assumed to be scalar, single-valued, and monotonic; increasing fields invariably yield reduced permeability and augmented reluctivity. This fact is of considerable theoretical importance, for it guarantees solvability and uniqueness for the methods presented in this chapter.

To attempt solution of Eqs. (1.03) or (1.04) by finite element methods, a suitable functional must first be defined. It would be tempting to assume that, in view of the similarity of (1.05) to the corresponding linear differential equations, a suitable functional for (1.04) would be

$$F(\mathbf{U})=\int_{\Omega}\frac{\mathbf{B}\cdot\mathbf{H}}{2}\,d\Omega-\int_{\Omega}\mathbf{J}\cdot\mathbf{U}\,d\Omega, \tag{1.06}$$

exactly as for the linear case; but this would not be quite correct. However, the integrand in the first term of (1.06) is readily spotted as being the stored energy density in the linear problem so that (1.06) may be written in the form

$$F(\mathbf{U})=\int W(\mathbf{U})\,d\Omega-\int\mathbf{J}\cdot\mathbf{U}\,d\Omega, \tag{1.07}$$

where $W(\mathbf{U})$ denotes the energy density associated with the trial solution $\mathbf{U}$. Happily, the energy density is not intrinsically tied to linear materials, so that (1.07) may be regarded as valid for nonlinear materials as well – provided the energy density is correctly taken for magnetic materials as

$$W = \int \mathbf{H} \cdot d\mathbf{B}. \tag{1.08}$$

In the vector potential problem, Eq. (1.04), one then has

$$\mathbf{B} = \nabla \times \mathbf{U} \tag{1.09}$$

and

$$\mathbf{H} = \nu \mathbf{B}. \tag{1.10}$$

The functional (1.07) corresponds to Eq. (6.07) in Chapter 2, examined in detail there.

It can be shown that, since the leading term in (1.07) actually represents energy, the conditions for existence and uniqueness of a mathematical solution are precisely those which are required for the existence of a unique stable state in a physical magnetic system: the magnetisation curves of all materials in the problem must be monotonic increasing, while their first derivatives are monotonic decreasing. Thus, for the purposes of this chapter, it will suffice to consider methods for minimising $F(\mathbf{U})$ as given by Eq. (1.07). Also, the two-dimensional approximation discussed above, with $\mathbf{U}$ an entirely z-directed quantity U, will be assumed.

2.   Minimisation over finite elements

The discretisation of functional (1.07) for the nonlinear case follows much the same techniques as were employed for linear problems. Let the problem region R be discretised into a set of nonoverlapping finite elements in the usual manner and let attention be concentrated on a single element. The potential A will be represented by the interpolative approximation

$$U = \sum_i U_i \alpha_i(x, y) \tag{2.01}$$

within a single element. With this substitution, $F(A)$ becomes an ordinary function of the finite number of finite element variables. The minimisation process required is thus simply

$$\frac{\partial F}{\partial U_i} = 0, \tag{2.02}$$

where the index i ranges over all unconstrained variables. Differentiating

Eq. (1.07) in order to minimise,

$$\int \left(\frac{\partial W}{\partial U_i} - J\alpha_i\right) d\Omega = 0. \tag{2.03}$$

From Eqs. (1.08) and (1.10) it follows that

$$\frac{\partial W}{\partial U_i} = \frac{\partial}{\partial U_i} \int_0^B \nu b \, db, \tag{2.04}$$

where b is a dummy variable of integration. A similar result holds for the scalar potential problem (1.03). In the following, it will be convenient to regard the reluctivity as a function of the square of the flux density, so that it is preferable to rewrite (2.04) as

$$\frac{\partial W}{\partial U_i} = \frac{1}{2} \frac{\partial}{\partial U_i} \int_0^{B^2} \nu(b^2) \, d(b^2), \tag{2.05}$$

which, by the usual chain rule of differentiation, becomes

$$\frac{\partial W}{\partial U_i} = \tfrac{1}{2}\nu(B^2) \frac{\partial}{\partial U_i} B^2. \tag{2.06}$$

Returning now to Eq. (2.03), which expresses the minimality requirement, let (2.06) be substituted. Since Eq. (1.09) relates the flux density **B** to the vector potential, Eq. (2.03) is readily brought into the matrix form

$$\mathbf{SU} = \mathbf{J}, \tag{2.07}$$

where **U** is the vector of nodal potential values, while **J** is a vector with entries

$$J_k = \int J\alpha_k \, d\Omega \tag{2.08}$$

and the element coefficient matrix **S** contains the entries

$$S_{ij} = \int \nu \left(\frac{\partial \alpha_i}{\partial x} \frac{\partial \alpha_j}{\partial x} + \frac{\partial \alpha_i}{\partial y} \frac{\partial \alpha_j}{\partial y}\right) d\Omega. \tag{2.09}$$

Of course, the reluctivity under the integral sign in (2.09) is not only a postition function but is also field dependent. Element matrix formation is therefore rather more complicated in the nonlinear than in the corresponding linear case, particularly since the material property may vary locally within any one element.

3.   Solution by simple iteration

In principle, any of the standard methods applicable to nonlinear equations may be employed for solving (2.07). A conceptually simple iterative method may be set up as follows. Let a set of reluctivity values be assumed, and let the resulting linear problem be solved. The potential vector **U** thus calculated will of course be wrong for the nonlinear problem. But **U** may be used to compute a new set of reluctivities, which can in turn be employed to calculate another approximation to **U**. The process can be begun by assuming initial (zero-field) reluctivity values everywhere, and can be terminated when two successive solutions agree within some prescribed tolerance. A suitable program flow chart is indicated in Fig. 5.1(a).

The simple algorithm of Fig. 5.1(a) is not always stably convergent, but it can be improved considerably by undercorrecting the reluctivities. That is to say, the reluctivities used during step $k+1$ are not those calculated from the approximate potential values at step k but, rather, a weighted combination of the potential values obtained at step k and step $k-1$:

$$\nu^{(k+1)} = \nu(p\mathbf{U}^{(k)} + (1-p)\mathbf{U}^{(k-1)}), \tag{3.01}$$

where the superscripts denote iteration steps. Here p is a fixed numerical parameter, $0 < p < 1$. Without undercorrection, the iterative process frequently oscillates over a wide range of reluctivity values. No extra memory is required by this change, since approximation $k-1$ for **U** can

Fig. 5.1. (a) A simple iterative algorithm. (b) Flow chart for a more reliable and possibly faster variant of the same method.

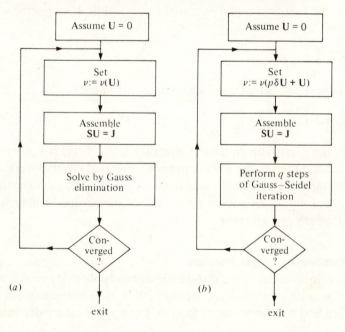

be overwritten by approximation $k + 1$; the required storage still amounts to two full vectors.

At each step of the iterative algorithm, a full set of simultaneous linear equations must be solved, of the form (2.07). Solution by standard elimination methods may seem wasteful for two reasons. First, a fairly good approximation to the solution is known at each step, possibly accurate to several significant figures; but no use is made of this prior information. Second, successive approximate solutions are all calculated to full precision even though the trailing figures are in most cases not significant at all. Rather than solve explicitly, it would therefore seem reasonable to employ an iterative method, such as the Gauss–Seidel iteration, for the simultaneous linear equations. Such methods can exploit the already existing approximate solution, and can also trade speed for accuracy by restricting the number of Gauss–Seidel iterations performed at each step of the iterative functional minimisation. A refined version of the iteration algorithm is shown in Fig. 5.1(b); numerous variants on it can be constructed by varying the number of steps q in the linear iteration between successive recalculations of reluctivity, and by choosing various different iterative methods for the linear equations. As a general rule, this class of methods tends to be memory-economic since it is never required to assemble the entire matrix $\mathbf{S}$. It is only necessary to form the product $\mathbf{SU}$ for given $\mathbf{U}$, and this requirement can be met by accumulating products on an element-by-element basis. Of course, the labour involved in reconstructing $\mathbf{S}$ at each main iterative step must still be performed, for the element matrix must be calculated for every element at every step. This amount of labour is formidable for complicated elements. However, in two-dimensional plane problems involving first-order triangular elements, the work is relatively minor.

Consider a triangular element in the x–y-plane and assume that all currents and vector potentials are purely z-directed. Let first-order triangular elements be used. Since the potential $\mathbf{A}$ varies linearly in each element, $\mathbf{B}$ must have a constant value throughout every element. To be consistent, the reluctivity must then also be constant in each element. Thus (2.09) becomes identical to its linear counterpart, so that it is very simple to evaluate. The only extra work required is the determination of the proper reluctivity value for each triangle.

Axisymmetric problems involving saturable materials are treated in much the same fashion as the corresponding problems with linear materials. It should be noted once again, however, that the vector and scalar Laplacian operators in cylindrical coordinates are in fact two different operators; hence the functionals for the vector and scalar potential problems are not identical. The differences, however, are similar to those encountered in linear problems, and will not be treated further here.

4.   A lifting magnet

To illustrate the manner of solving nonlinear problems, consider the simple lifting magnet shown in Fig. 5.2(*a*), It will be assumed that the magnet structure extends into the paper relatively far, so that the problem may be considered to be two-dimensional. Of course, the structure is in reality not only three-dimensional, but is further complicated by being unbounded. However, the major interest usually centres on the local fields near the magnet poles. Hence, taking the problem region to be bounded at some relatively great distance, as well as treating it in two dimensions, are reasonable assumptions. In Fig. 5.2(*a*), the magnet is therefore encased in a rectangular box whose sides are assumed to be at zero magnetic vector potential. Only half the problem need be analysed because the left edge of Fig. 5.2(*a*) represents a symmetry plane.

Fig. 5.2. (*b*) Outline drawing of a lifting magnet. (*b*) Finite element mesh used for analysis of the magnet. (Screen photographs taken from a computer-aided analysis and design system.)

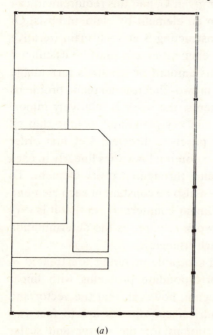

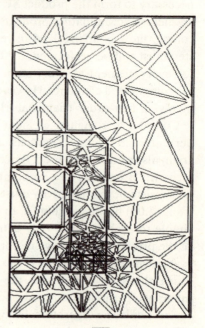

(*a*)   (*b*)

As in the various linear problems treated in Chapter 1, the problem region is first modelled by an assembly of finite elements. The choice of element size and shape should take account of the probable stored energy density in any given part of the region; where the energy density is high, and varies rapidly, the elements should be smallest. However, the nonlinearity of the material prevents energy densities from rising quite so high as would be the case for linear materials because the iron permeability will fall as the flux density rises. Thus, a somewhat more even element subdivision is adequate for the nonlinear case. Fig. 5.2(*b*) shows a decomposition of the problem region of Fig. 5.2(*a*). Near the magnet pole, many quite small elements are employed; not all of these are distinctly visible in the diagram. The mesh shown includes some 275 elements.

A flux plot of a solution is shown in Fig. 5.3(*a*). The flux distribution in the iron near the air gap and particularly at the corners of the magnet iron is relatively uniform, showing some, but not very strong, iron saturation. The iron plate being lifted, being much thinner than the lifting magnet itself, is strongly saturated. A flux line count shows that the leakage flux, which avoids the plate entirely, amounts to about 20 per cent. This figure is of interest to designers, for it is one of several useful indices of design effectiveness.

While designers often desire overall flux line plots, to assess global characteristics of the solution, more importance usually attaches to the

Fig. 5.3. (*a*) **Flux distribution in the lifting magnet, with a current density of 100 amperes per square centimetre in the exciting coils.** (*b*) **Enlarged plot of the air-gap region of the magnet.**

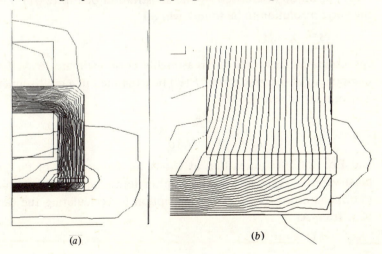

(*a*)   (*b*)

air-gap fields. Thus Fig. 5.3(*b*) shows an enlarged portion of the air gap, which permits closer examination of the field distribution in the magnet pole as well as in the plate being lifted.

5.   The Newton iteration process

The simple iteration method given above converges rather slowly and consumes considerable quantities of computer time. Further, there is no known certain method for estimating the amount of reluctivity undercorrection that will guarantee stability without slowing the process unduly. It is therefore preferable to examine the Newton iteration process, which is unconditionally stable. It is very much more rapidly convergent than simple iteration and, surprisingly enough, requires little if any additional computer memory. Its convergence rate is theoretically quadratic near the solution point, i.e., the number of significant digits in each iterative trial solution should be approximately double the number of significant digits in the preceding one. Thus, if the first step of iteration achieves an approximate solution correct to within a factor of two (correct to one significant binary digit), the second step will produce two correct bits, the third one four, the next eight, and so on. Proceeding further, five or six quadratically convergent steps should yield the full precision available on a conventional 32-bit digital computer. Truly quadratic convergence is in fact only obtained very near the solution. Nevertheless, more than seven or eight Newton steps are not very often required for precision exceeding the level physically justifiable.

The Newton process may be set up as follows. Let the energy functional (1.07) again be discretised by the substitution of Eq. (2.01). Let **A** be the correct solution to be found, while **U**,

$$\mathbf{U} = \mathbf{A} - \delta\mathbf{U}, \tag{5.01}$$

represents an incorrect but reasonably close estimate to **A**. Let each component of the gradient of $F(\mathbf{U})$ be expanded in a multi-dimensional Taylor's series near **U**:

$$\frac{\partial F}{\partial U_i} = \frac{\partial F}{\partial U_i}\bigg|_U + \sum_j \frac{\partial^2 F}{\partial U_i \, \partial U_j}\bigg|_U \delta U_j + \dots . \tag{5.02}$$

But Eq. (2.02) requires that at $\mathbf{U} = \mathbf{A}$ all components of the gradient must vanish. Neglecting Taylor's series terms beyond the second, Eq. (5.02) therefore furnishes a prescription for calculating the deviation of **U** from **A**,

$$\delta\mathbf{U} = -\mathbf{P}^{-1}\mathbf{V}, \tag{5.03}$$

where $\mathbf{P}$ is the Jacobian matrix of the Newton iteration (the *Hessian* matrix of F),

$$P_{ij} = \frac{\partial^2 F}{\partial U_i \, \partial U_j},$$
(5.04)

while $\mathbf{V}$ represents the gradient of $F(\mathbf{U})$ at $\mathbf{U}$:

$$V_j = \frac{\partial F}{\partial U_j}.$$
(5.05)

An iterative method is now constructed by assuming some set of potentials $\mathbf{U}$ to start, and calculating its apparent difference from $\mathbf{A}$, using Eq. (5.03). This computed difference (usually called the displacement) is then added to the initially estimated potentials. Since third- and higher-order derivatives of the Taylor's series (5.02) were discarded, the displacement as computed by Eq. (5.03) is only approximate. It is exact if $F(\mathbf{U})$ is exactly quadratic, so that no third- or higher-order derivatives exist; indeed, this is the case for linear problems, and the iteration converges in one step. For nonlinear problems, the computed displacement is simply added to $\mathbf{U}$ to form a better estimate, and the entire process is repeated. Thus, the iterative prescription

$$\mathbf{U}^{(k+1)} = \mathbf{U}^{(k)} - [\mathbf{P}^{(k)}]^{-1}\mathbf{V}^{(k)}$$
(5.06)

produces the sequence of Newton iterates converging to $\mathbf{A}$.

Successful implementation of the Newton process (5.06) requires evaluation of the derivatives (5.04) and (5.05). Formal differentiation of $F(\mathbf{U})$ yields

$$\frac{\partial^2 F}{\partial U_i \, \partial U_j} = \int \frac{\partial^2 W}{\partial U_i \, \partial U_j} \, d\Omega.$$
(5.07)

On differentiating Eq. (2.06) the integrand in turn becomes

$$\frac{\partial^2 W}{\partial U_i \, \partial U_j} = \frac{\nu}{2} \frac{\partial^2 (B^2)}{\partial U_i \, \partial U_j} + \frac{1}{2} \frac{d\nu}{dB^2} \frac{\partial B^2}{\partial U_i} \frac{\partial B^2}{\partial U_j}.$$
(5.08)

In accordance with Eqs. (1.09) and (2.01),

$$B^2 = \sum_m \sum_n (\nabla \alpha_m \cdot \nabla \alpha_n) U_m U_n.$$
(5.09)

The derivatives in (5.08) may thus be explicitly evaluated, yielding

$$\frac{\partial^2 W}{\partial U_i \, \partial U_j} = \nu (\nabla \alpha_i \cdot \nabla \alpha_j) + 2 \frac{d\nu}{d(B^2)}$$

$$\times \sum_m \sum_n (\nabla \alpha_m \cdot \nabla \alpha_i)(\nabla \alpha_n \cdot \nabla \alpha_j) U_m U_n.$$
(5.10)

It is interesting to note that the first right-hand term in Eq. (5.10) is exactly what would have been expected for a linear problem. The second and more complicated term only appears in the nonlinear case. Thus, somewhat more calculation will be required to construct the coefficient matrix **P** for a nonlinear problem than would have been required for the coefficient matrix **S** of a similar linear problem. The matrix structure, however, is exactly the same in both cases. Thus the nonlinear problem requires a little more computation, but imposes the same memory requirements.

6.   First-order, triangular, Newton elements

In electric machine analysis, under the assumption of infinite axial machine length, first-order triangular elements have proved very popular. Their usefulness is due partly to their geometric flexibility which permits easy modelling of complicated electric machine cross-sections, and partly to their relative simplicity in computation.

To develop first-order triangular elements, let the Jacobian matrix of Eq. (5.04) be written out in detail. It assumes the form

$$P_{ij} = \int \nu \nabla \alpha_i \cdot \nabla \alpha_j \, d\Omega$$

$$+ 2 \int \frac{d\nu}{d(B^2)} \sum_m \sum_n (\nabla \alpha_m \cdot \nabla \alpha_i)(\nabla \alpha_n \cdot \nabla \alpha_j) U_m U_n \, d\Omega. \qquad (6.01)$$

On a first-order triangular element, the flux density and therefore the reluctivity are constant everywhere. The first integral in Eq. (6.01) may thus be recognised as being simply the nonlinear **S** of Eq. (2.09). Further, since the first-order element interpolation functions are linear, their gradients must be constants. Hence

$$P_{ij} = S_{ij} + 2 \frac{d\nu}{d(B^2)} \Delta \sum_m \sum_n U_m (\nabla \alpha_m \cdot \nabla \alpha_i)(\nabla \alpha_n \cdot \nabla \alpha_j) U_n, \qquad (6.02)$$

where Δ denotes the element area. To simplify notation, let **S'** denote the matrix **S** obtained with unity reluctivity,

$$S_{ij} = \nu S'_{ij} \qquad (6.03)$$

and let **E** be the auxiliary vector

$$E_k = \sum_m S'_{km} U_m. \qquad (6.04)$$

With these notational alterations, Eq. (6.02) can easily be brought into the form

$$P_{ij} = \nu S'_{ij} + \frac{2}{\Delta} \frac{d\nu}{d(B^2)} E_i E_j. \qquad (6.05)$$

The second term on the right, it may be noted, is merely the outer product of the auxiliary vectors $\mathbf{E}$, hence it is computationally very cheap.

The remaining terms in the equations that describe the Newton process with first-order triangular elements are easily added, indeed they were largely worked out for the simple iterative process above. The residual vector (5.05) may be written

$$V_i = \int \left. \frac{\partial W}{\partial U_i} \right|_U d\Omega - J_i, \tag{6.06}$$

where $\mathbf{J}$ is the vector defined by Eq. (2.08). Differentiating in accordance with Eq. (2.06), and using the notations defined immediately above,

$$\int \frac{\partial W}{\partial U_i} d\Omega = \nu \sum_m S'_{im} U_m = \nu E_i. \tag{6.07}$$

The entire iterative process thus runs as shown in Fig. 5.4.

An operation count shows that to form the Jacobian matrix of Eq. (6.05) and the residual vector of Eq. (6.06) requires a total of twenty-eight multiplications per finite element, plus whatever arithmetic may be needed to determine the values of reluctivity and its derivative. This computing time is inherently short; furthermore, for many elements the

Fig. 5.4. The Newton algorithm, stably covergent for magnetic field problems wherever the magnetisation characteristics are monotonic.

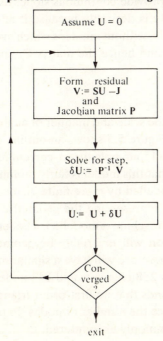

Assume $U = 0$

Form residual
$V := SU - J$
and
Jacobian matrix P

Solve for step,
$\delta U := P^{-1} V$

$U := U + \delta U$

Con-
verged
?

exit

total time increases only linearly with element number. On the other hand, even the most efficient sparse-matrix equation-solving algorithms have solution times rising more rapidly than linearly with node number. In consequence, matrix set-up time equals solution time for rather small numbers of elements (of the order of 50 or fewer), while in practical analysis problems involving many hundreds of elements, matrix formation only requires about 5–10 per cent of total computing time. Thus total computing times are sensitive to the choice of equation-solving method for finding the Newton correction, Eq. (5.03).

It may be concluded from the above rough estimates that the computing time required for finding reluctivities from given potential values is of importance. Obviously, the material properties should be characterised by data so arranged as to yield reluctivity values quickly, given the square of flux density. The classical *B–H*-curve is decidedly not the best choice here, and curves relating reluctivity to the squared flux density directly are commonly used. For economy it is best to avoid modelling the reluctivity curve by transcendental functions, which usually entail a substantial computing cost. For example, an exponential function typically entails a computing cost equivalent to ten multiply-and-add operations. Good choices are cubic splines with equispaced knots, or cubic Hermite interpolation polynomials. Both require only three multiplications for evaluation, and both provide continuous slopes as well as function continuity. Slope continuity is desirable because it allows the Newton process to converge smoothly, without missteps which may occur if the first derivative of reluctivity, and hence the matrix **P**, undergoes sharp changes.

7.   Analysis of a DC machine

An example may illustrate the scale and manner of analysis now used in various industrial settings. Figure 5.5 shows an outline drawing and a finite element model for a small four-pole direct-current machine. It should be evident on inspection that this model is fairly minimal; very few portions of the machine are modelled by more finite elements than are actually necessary for correctly representing the geometric shapes. Only one pole pitch of the four-pole machine is modelled. The remaining portion of the machine cross-section will invariably be geometrically similar, similar in excitations, and therefore will have a similar magnetic field solution. This model comprises 258 elements and 143 node points. It can be shown on theoretical grounds that no first-order triangulation can contain more elements than twice the number of nodes. In practical analyses, ratios of about 1·8 are commonly encountered.

A solution based on this model, computed for full rated field current but no other winding currents, is shown in Fig. 5.6. The boundary conditions imposed to obtain this solution are fairly simple. The outer periphery of the machine may be taken to be a flux line, setting $A = 0$ all along it; and the centrepoint may be supposed to have $A = 0$ also, for it must surely lie on the same flux line. When the machine is running unloaded, the rotor current and interpole current are absent, so that the field winding provides the only excitation. In these circumstances, the pole axis and the interpolar axis may be taken to be symmetry lines; a flux line must in general run along the pole axis, while the interpole axis must be orthogonal to all flux lines. However, the fact that the rotor structure is toothed in Fig. 5.5 may result in there existing no clearcut symmetry line. Flux lines will therefore generally cross the pole axis in some unspecified manner. But it is clear that the potential A must have the same set of values (except for reversal of sign) at corresponding points along the interpolar axes as well as the pole axes. In other words, if i and j are geometrically corresponding nodes on the axes of adjacent

Fig. 5.5. Finite element model of a moderate-sized direct-current machine. The model comprises 258 elements per half pole pitch.

poles (the vertical and horizontal axes in Fig. 5.5, for example), only one of the two potentials is independently specifiable:

$$A_i = -A_j. \tag{7.01}$$

This constraint is essentially similar to the constraints encountered in element interconnection; in fact it *is* an element interconnection constraint. The only novelty is perhaps that the elements being interconnected are not located in geometrically adjacent positions, and that a negative sign appears in the constraint matrix **C**. In this situation, a full pole pitch must be modelled for solution; there is no symmetry condition that would permit working with half a pole pitch only. In the model of Fig. 5.5, there are 24 constraint equations arising from boundary conditions, i.e., the connection matrix **C** has 774 rows and 119 columns, so that the magnetic field is computed by solving 119 simultaneous equations at every step of the Newton iteration.

The flux linking any given contour (such as, for example, the conductor in a given winding) may be calculated (see Chapter 2, Section 2.4) as

Fig. 5.6. No-load flux distribution in the four-pole machine shown in Fig. 5.5. The flux lines exhibit symmetry about both pole and interpole axis, as might be expected.

the line integral of vector potential around that contour,

$$\varphi = \oint \mathbf{A} \cdot \mathbf{dl}. \tag{7.02}$$

Since end effects are neglected here, only longitudinally directed portions of the contour contribute to the integral, the flux linkages per unit length can be calculated by adding all the vector potentials at coil sides, being careful to multiply the potential values by -1 wherever the coil-winding direction coincides with the z-direction, and by $+1$ where the winding direction is colinear with the z-axis. Thus

$$\varphi = \sum_{i=1}^{N} w_i [A_i]. \tag{7.03}$$

Here the coefficients w take on values of $+1$ or -1 according to winding direction, while N is the number of turns in the winding. The index i ranges over all the elements belonging to the winding for which the flux linkages are to be calculated, while the bracketed quantity on the right is the average vector potential in element i. If it is assumed that the winding is made of conductors much smaller in diameter than the element size, successive winding turns will occupy all possible positions in each element, hence the need for averaging. If desired, an approximation to the generated voltage waveshape in the winding may also be found by this method without recomputing the field distribution. It suffices to recalculate the flux linkages, Eq. (7.03), for successive positions of the rotor or, what is equivalent, recalculating the summation (7.03) with all winding-direction coefficients shifted one slot.

A quite minimal finite element model such as that in Fig. 5.5 is usually sufficient for computing machine-performance indices which depend on overall averages or weighted averages of local potential values. Flux linkages, generated voltages, stored energies, and all terminal parameters belong to this class. Models of this level of sophistication may be expected to produce terminal voltages with an accuracy of the order of a few per cent at worst. This accuracy should not be surprising, for the classical, simple magnetic circuit approach to generated voltage estimation may be expected to come within 5–6 per cent of experiment; and it may itself be regarded as a very crude finite element method! For purposes of calculating terminal parameters, first-order triangular element solutions are quite good enough if only sufficient elements are used to model the machine shape accurately; very fine subdivision throughout the entire machine is usually quite unnecessary. Naturally, if local details of flux distributions are of interest, locally refined meshes should be employed. For example, if tooth-tip flux density distributions are considered to be of importance, very considerably finer subdivision should be used for the tooth to be investigated and probably for its neighbours to either side as well.

When running under load, the flux distribution in the machine is no longer symmetric about the interpolar axis, as may be seen from Fig. 5.7 which exhibits the calculated flux distribution under load. In this case, a full pole pitch of the machine must be analysed; furthermore, boundary conditions become somewhat more complicated than in the no-load case. It is still valid to hold the outside rim of the machine to be a flux line, $A = 0$, and to insist that this flux line must pass through the machine centreline. However, while in the no-load case symmetry requires that this flux line follow the pole axis in radial direction, it is clear from Fig. 5.7 that no particular position can be assigned to it in the loaded case. What can be said, however, is that all phenomena in the machine must be periodic. Hence, all vector potentials along a radial

Fig. 5.7. **Flux distribution under load in the four-pole machine shown in Fig. 5.5. The flux lines obey the required periodicity rules at the polar and interpolar axes; but the axes are not symmetry lines.**

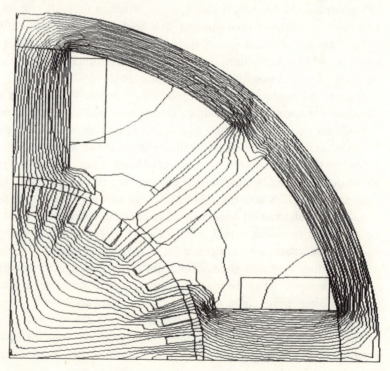

line must be exactly the negatives of the corresponding vector potentials along another radial line half a pole pitch away. For example, the vector potential values along the vertical pole axis in Fig. 5.7 must be the negatives of the vector potentials along the horizontal pole axis. This constraint is amenable to exactly the same treatment as all the usual boundary conditions. For simplicity, consider the mesh shown in Fig. 5.6(*a*) where potentials at nodes 5 and 6 will be required to equal the negatives of potentials at points 1 and 2. The corresponding disjoint element assemblage is shown in Fig. 5.8(*b*). Clearly, in the assembled problem there will be only four independent potential values. Thus the vector of twelve disjoint potentials must be related to the connected potentials by the matrix transformation

$$\mathbf{A}_{\text{dis}} = \mathbf{C}\mathbf{A}_{\text{con}}, \tag{7.04}$$

where the connection matrix $\mathbf{C}$ is the transpose of

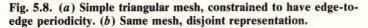

$$\mathbf{C}^{\mathrm{T}} = \begin{bmatrix} 1 & & & -1 & & & -1 & \\ & 1 & & & -1 & 1 & & \\ & & 1 & & & & 1 & 1 & \\ & & & 1 & & & & 1 & & 1 \end{bmatrix}. \tag{7.05}$$

It follows that the periodicity requirement implies a process of matrix assembly which is in principle similar to any other, but involves subtraction of some entries since the connection matrix in this case involves negative as well as positive elements.

Fig. 5.8. (*a*) Simple triangular mesh, constrained to have edge-to-edge periodicity. (*b*) Same mesh, disjoint representation.

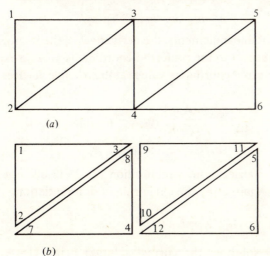

8.   Anisotropic materials

In many applications anisotropic materials are deliberately employed, as, for example, grain-oriented steels in transformer cores. Such materials have one reluctivity value in a preferred direction p, and a different value in the orthogonal direction q. The arguments earlier applied to general fields, Eq. (1.05)–(1.10), remain valid, and the functional to be minimised remains $F(u)$ of Eq. (1.07), provided the reluctivity is treated as a tensor quantity. The difficulty in practical implementation lies in the fact that the reluctivity, and hence the stored energy density W, both depend not only on the magnitude of the flux density vector but also on its direction. In practice, it may be expected that curves of reluctivity against flux density are obtainable for the two principal directions p and q but only rarely for intermediate directions. An assumption which usually lies fairly close to the truth (the error rarely exceeds a few percentage points) is to suppose that loci of constant reluctivity form ellipses in the B_p–B_q-plane. In algebraic terms, this statement is equivalent to assuming that

$$W = W(B_p^2 + r^2 B_q^2), \tag{8.01}$$

r being the ratio of semiaxes. Let it be desired to construct triangular elements. In terms of the usual triangle area coordinates, the argument in (8.01) may be expressed as

$$B_p^2 + r^2 B_q^2 = \left(\frac{\partial A}{\partial q}\right)^2 + r^2 \left(\frac{\partial A}{\partial p}\right)^2 \tag{8.02}$$

or, rewriting,

$$B_p^2 + r^2 B_q^2 = \sum_i \sum_j \frac{\partial A}{\partial \zeta_i} \frac{\partial A}{\partial \zeta_j} \left(\frac{\partial \zeta_i}{\partial p} \frac{\partial \zeta_j}{\partial p} + r^2 \frac{\partial \zeta_i}{\partial q} \frac{\partial \zeta_j}{\partial q}\right). \tag{8.03}$$

Just as in the case of linear elements, the derivatives of the homogeneous coordinates with respect to the preferred coordinates may be expressed in terms of the preferred coordinate values at the triangle vertices. There results

$$B_p^2 + r^2 B_q^2 = \frac{1}{4\Delta^2} \sum_i \sum_j \frac{\partial A}{\partial \zeta_i} \frac{\partial A}{\partial \zeta_j} [(q_{i-1} - q_{i+1})(q_{j-1} - q_{j+1})$$
$$+ r^2(p_{i+1} - p_{i-1})(p_{j+1} - p_{j-1})]. \tag{8.04}$$

To form element matrices, the minimisation (2.02)–(2.03) next needs to be carried out. Again using the chain rule of differentiation,

$$\frac{\partial W}{\partial U_i} = \frac{\partial W}{\partial b^2} \frac{\partial b^2}{\partial U_i}, \qquad b^2 = B_p^2 + r^2 B_q^2. \tag{8.05}$$

Here $\mathbf{U}$ is a trial solution, the correct solution being $\mathbf{U} = \mathbf{A}$. Using interpolatory approximating functions, as in (2.01), substitution yields

$$\frac{\partial W}{\partial U_k} = \frac{1}{2\Delta^2} \sum_l U_l \sum_i \sum_j \frac{dW}{dB^2} \frac{\partial \alpha_l}{\partial \zeta_i} \frac{\partial \alpha_k}{\partial \zeta_j}$$
$$\times [(q_{i-1} - q_{i+1})(q_{j-1} - q_{j+1}) + r^2(p_{i+1} - p_{i-1})(p_{j+1} - p_{j-1})].$$
(8.06)

The indices i and j are of course taken to be cyclic modulo 3.

The matrix S for the finite element method may be obtained by integration of Eq. (8.05). Analytic integration is only possible in relatively simple cases because W is in general some relatively complicated function of position. To construct a Newton method, the chain rule of differentiation may again be applied so as to shift differentiations from the potentials to the flux density squared. There results

$$\frac{\partial^2 W}{\partial U_k \partial U_m} = \frac{d^2 W}{d(b^2)^2} \frac{\partial b^2}{\partial U_k} \frac{\partial b^2}{\partial U_m} + \frac{dW}{dB^2} \frac{\partial^2 (b^2)}{\partial U_k \partial U_m},$$
(8.07)

where the derivatives are evaluated exactly as for (8.05). Once more, analytic differentiation and integration are very difficult except for first-order triangles.

Methods of this type have been used extensively for the study of flux distribution and rotational flux densities in polyphase multi-limbed transformer cores using grain-oriented steel. In such cases, the preferred direction p is the direction of rolling of the steel sheet. The semiaxis ratio r above may be as high as 10, so that an analysis based on assumed isotropy does not yield good results.

9.   Readings

Nonlinear problems arise in many areas of continuum electrophysics, and finite element methods have been applied to a large variety of specific cases. Three main areas dominate the work to date: ferromagnetic nonlinear materials, semiconductors, and plasmas. While details differ considerably, the basic theory remains essentially similar to that applicable elsewhere in mathematical physics. The book by Oden (1972) outlines the applied mathematics underlying the entire area. A more recent survey of the techniques as applied to electrical engineering problems will be found in Chari & Silvester (1981).

The ferromagnetic materials problem was the first to be tackled by finite elements. To date, it remains the best-researched and best-established, with extensively published theory as well as substantial quantities of applicational examples available in the literature. The first papers dealing with this class of problems were those of Chari & Silvester

(1971), which defined a methodology for static fields. The methods, as set out in this chapter, largely follow the approach of Silvester, Cabayan & Browne (1973). Numerous advances since that time have widened the scope of the method, e.g., the treatment of time-varying processes as exemplified by Hanalla & Macdonald (1980), or the use of curvilinear elements by Silvester & Rafinejad (1974).

Semiconductor modelling work follows techniques similar in kind, but perhaps a little more complicated in detail. A good survey of the earlier work in the area will be found in Barnes & Lomax (1977). In this area, as in magnetics, techniques are now sufficiently advanced to allow construction of commercially available, widely-usable program packages, such as reported by Buturla, Cottrell, Grossman & Salsburg (1981).

The solution of Maxwell's equations in plasmas has applications in both the atomic energy (plasma containment) and radio propagation areas. Relatively less has been accomplished in that area by finite elements; the article by Tran & Troyon (1978) may provide a sample.

References

Barnes, J. J. & Lomax, R. J. (1977). 'Finite element methods in semiconductor device simulation', *Institute of Electrical and Electronics Engineers Transactions on Electron Devices*, **ED-24**, 1082-9.

Buturla, E. M., Cottrell, P. E., Grossman, B. M. & Salsburg, K. A. (1981). 'Finite element analysis of semiconductor devices: the FIELDAY prorgram', *IBM Journal of Research and Development*, **25**, 218-31.

Chari, M. V. K. & Silvester, P. (1971), 'Analysis of turboalternator magnetic fields by finite elements', *Institute of Electrical and Electronics Engineers Transactions on Power Apparatus and Systems*, **PAS-90**, 454-64.

Chari, M. V. K. & Silvester, P. P. (eds.) (1981). *Finite Elements for Electrical and Magnetic Field Problems*. Chichester: John Wiley.

Hanalla, A. Y. & MacDonald, D. C. (1980). 'Sudden 3-phase short-circuit characteristics of turbine generators from design data using electromagnetic field calculations', *Institution of Electrical Engineers Proceedings*, vol. 127, part C, pp. 213-20.

Oden, J. T. (1972), *Finite Elements of Nonlinear Continua*. New York: McGraw-Hill.

Silvester, P., Cabayan, H. & Browne, B. T. (1973). 'Efficient techniques for finite element analysis of electric machines', *Institute of Electrical and Electronics Engineers Transactions on Power Apparatus and Systems*, **PAS-92**, 1274-81.

Silvester, P. & Rafinejad, P. (1974). 'Curvilinear finite elements for two-dimensional saturable magnetic fields', *Institute of Electrical and Electronics Engineers Transactions on Power Apparatus and Systems*, **PAS-93**, 1861-70.

Tran, T. M. & Troyon, F. (1978). 'Finite element calculation of the anomalous skin effect in a homogeneous unmagnetized cylindrical plasma column', *Computer Physics Communications*, **16**, 515-6.

6
Three-dimensional problems

1.   Introduction

Some real-life situations have a symmetry which allows a description to be made in terms of two, not three, independent space coordinates. Solids of revolution with axial or circumferential field excitation fall exactly into this category, and long cylindrical geometries with axial excitation are also well approximated by two-dimensional models. However, many practical arrangements have no such symmetry and require specification by three independent coordinates with a full three-dimensional treatment. Here, in this book, methods of analysis involving the division of space into finite elements, defined by vertex nodes, are being considered. Values of engineering parameters at the nodal points are assumed to be sufficient to represent the solution of a problem. It may be accepted that achieving a specified accuracy by a given method is dependent upon the linear spacing between nodes. Thus, equivalent two- and three-dimensional problems will entail the employment of n^2 and n^3 nodes, respectively, where n is some number appropriate to the accuracy of solution required. The computer manipulation of such a discretisation in the three-dimensional case will require correspondingly greater central processor time and storage, to the extent that a problem which was manageable in two dimensions might now become impossible to handle. For problems involving vector field functions the penalty incurred may in fact be greater than that associated merely with changing from n^2 to n^3 variables. This is well illustrated by the example of magnetostatic field calculations employing the equation

$$\nabla \times (\nu \nabla \times \mathbf{A}) = \mathbf{J} \tag{1.01}$$

in vector potential $\mathbf{A}$, which defines magnetic field through the operation

$$\mathbf{B} = \nabla \times \mathbf{A}. \tag{1.02}$$

This example was discussed in Chapter 3, Section 5. In the two-dimensional case with translationally-symmetric geometry, the result of specifying axially-directed current distributions,

$$\mathbf{J} = (0, 0, J(x, y)), \tag{1.03}$$

is a two-dimensional vector potential also wholly directed axially:

$$\mathbf{A} = (0, 0, A(x, y)). \tag{1.04}$$

Note that the vector given by Eq. (1.04) identically satisfies the gauge condition

$$\nabla \cdot \mathbf{A} = 0. \tag{1.05}$$

The exercise has thus been effectively reduced to one in a single, two-dimensional scalar variable, governed by the scalar equation

$$\nabla \cdot (\nu \nabla A) = 0. \tag{1.06}$$

The vector nature of the problem does not enter until $A(x, y)$ has been determined. The derivation of the vector field $\mathbf{B}$ through Eq. (1.02) is not an expensive computer operation.

If the three-dimensional case is considered, however, it appears that all three components of $\mathbf{A}$ must be determined at each node, so the n^2 number of variables in two dimensions now becomes $3n^3$. A gauge condition, such as Eq. (1.05), can of course be specified, in which case it seems possible that some reduction in the arithmetic required might occur if this condition were exploited. However, the advantages of a symmetric set of variables and the simpler boundary constraints associated with the full vector approach seem, at the present time, to outweigh any economy obtainable by this means.

In the general electromagnetics case with time variation, the full set of variables is represented by $\mathbf{A}$ together with the scalar potential V (see Chapter 2, Section 2). Only then are the field variables

$$\mathbf{E} = -\nabla V - \frac{\partial \mathbf{A}}{\partial t}, \tag{1.07}$$

$$\mathbf{B} = \nabla \mathbf{A} \tag{1.08}$$

adequately defined. However, a gauge assumption, typically

$$\nabla \cdot \mathbf{A} = -\mu \varepsilon \frac{\partial V}{\partial t} \tag{1.09}$$

may be made. This allows the time-varying problem, at first sight apparently requiring *four* independent scalar variables, in fact to be worked in terms of three. These might be, for instance, A_x, A_y and A_z with V derived as a simple time integral from $\nabla \cdot \mathbf{A}$.

2.   Tetrahedral scalar elements

The extra difficulties arising in three-dimensional problems associated with a vector working variable have been discussed. Nevertheless, there are many situations where calculations can be performed entirely in one scalar variable and it is appropriate to examine such problems first. The time-variant, scalar, three-dimensional problem is typified by the case of determining the electrostatic potential within a volume Ω bounded by electrodes at specified potentials. Here, the governing equation in potential u is

$$\nabla \cdot (\varepsilon \nabla u) = -\rho \tag{2.01}$$

for a medium with space-varying permittivity ε and charge distribution ρ (see Chapter 2, Section 3). The determination of the resonant frequency of an acoustic chamber represents an extension of the scalar problem to include harmonic time variation. In this case the equation is

$$\nabla^2 p + k^2 p = 0, \tag{2.02}$$

where p is the pressure perturbation at frequency ω from its static value whilst, with c corresponding to the velocity of sound in the cavity, $k = \omega/c$ represents eigenvalues which must be calculated in order to determine resonant frequencies (see Chapter 2, Section 8).

Such typical cases were observed in Chapter 2, Section 3 to be covered by the inhomogeneous Helmholtz equation

$$\nabla \cdot (p \nabla u) + k^2 u = g, \tag{2.03}$$

subject to the Dirichlet or homogeneous Neumann conditions. (The material property p in Eq. (2.03) should not be confused with the pressure variable in Eq. (2.02).) It was shown that the functional

$$F(U) = \tfrac{1}{2} \int_\Omega [p(\nabla U)^2 - k^2 U^2 + 2gU] \, d\Omega, \tag{2.04}$$

with U constrained to conform to the Dirichlet boundary conditions but otherwise free to vary, is stationary about the true solution

$$U = u. \tag{2.05}$$

Following the two-dimensional variational analysis given for the inhomogeneous Helmholtz equation in terms of triangles (Chapter 3), an exactly analogous procedure in three dimensions, first given by Silvester (1972), is followed with respect to tetrahedral elements. It will be shown that a matrix equation may be set up to determine the vector array **U** of potentials at the nodes of a lattice built up from tetrahedra. The analysis is once more initially worked in terms of a single element.

The potentials are approximate because of the limitations imposed by the assumption of U being a polynomial function of the space variables within the tetrahedron. Subsequently, the procedure by which the results from a single tetrahedron can be applied to an ensemble of connected elements is discussed.

2.1.   *Three-dimensional homogeneous coordinates*

Homogeneous coordinates are used as in the two-dimensional case, allowing the essential properties of the simplex element (see Fig. 6.1) due to its tetrahedral form to be separated from its particular shape, size and position in space. The tetrahedral coordinates are defined in the following way: let P be a point within the tetrahedron having Cartesian coordinates (x, y, z). Suppose the volumes of the four smaller tetrahedra, each given by three of the original four vertices (x_i, y_i, z_i), $i = 1, 2, 3, 4$, and P, are V_1, V_2, V_3 and V_4. Then the homogeneous coordinates of P are defined by $(\zeta_1, \zeta_2, \zeta_3, \zeta_4)$ where, with V being the total volume of the tetrahedron,

$$\zeta_1 = V_1/V, \tag{2.06}$$

$$\zeta_2 = V_2/V, \tag{2.07}$$

and so forth. Clearly, the four coordinates are not linearly independent but satisfy

$$\sum_{i=1}^{4} \zeta_i = 1. \tag{2.08}$$

It is noted that

$$V = \frac{1}{3!} \begin{vmatrix} 1 & x_1 & y_1 & z_1 \\ 1 & x_2 & y_2 & z_2 \\ 1 & x_3 & y_3 & z_3 \\ 1 & x_4 & y_4 & z_4 \end{vmatrix} \tag{2.09}$$

Fig. 6.1. Tetrahedron simplex three-dimensional finite element.

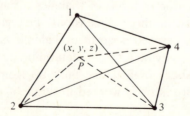

and that

$$V_1 = \frac{1}{3!} \begin{vmatrix} 1 & x & x & z \\ 1 & x_2 & y_2 & z_2 \\ 1 & x_3 & y_3 & z_3 \\ 1 & x_4 & y_4 & z_4 \end{vmatrix}, \tag{2.10}$$

$$V_2 = \frac{1}{3!} \begin{vmatrix} 1 & x_1 & y_1 & z_1 \\ 1 & x & y & z \\ 1 & x_3 & y_3 & z_3 \\ 1 & x_4 & y_4 & z_4 \end{vmatrix}, \tag{2.11}$$

with similar expressions for V_3 and V_4. Hence it is seen that the relationship

$$\zeta_i = \frac{a_i + b_i x + c_i y + d_i z}{3!V} \tag{2.12}$$

may be written in which a_i, b_i, c_i and d_i are appropriate cofactors picked from determinants like those displayed in Eqs. (2.10) and (2.11). Thus the coordinate system $(\zeta_1, \zeta_2, \zeta_3, \zeta_4)$ is defined explicitly in terms of the tetrahedron vertex Cartesian coordinates.

2.2 Interpolation polynomials

Following the procedure adopted in the two-dimensional cases of Chapter 3, interpolation polynomials

$$\alpha_{ijkl} = R_i(n_0, \zeta_1) R_j(n_0, \zeta_2) R_k(n_0, \zeta_3) R_l(n_0, \zeta_4), \tag{2.13}$$

with $i + j + k + l = n_0$, are set up, where, as in Eq. (3.01) of Chapter 3,

$$R_m(n_0, \zeta) = \frac{1}{m!} \prod_{k=0}^{m-1} (n_0 \zeta - k) \quad m > 0, \qquad R_0(n_0, \zeta) = 1. \tag{2.14}$$

It is readily verified that $\alpha_{ijkl}(\zeta_1, \zeta_2, \zeta_3, \zeta_4)$ is interpolatory and a polynomial of degree n_0. It vanishes at the regularly spaced set of points $\zeta_1 = p/n_0$, $\zeta_2 = q/n_0$, $\zeta_3 = r/n_0$, $\zeta_4 = s/n_0$ for integral values of p, q, r and s with $p + q + r + s = n_0$ excepting the case $p = i$, $q = j$, $r = k$, $s = l$, when

$$\alpha_{ijkl}(i/n_0, j/n_0, k/n_0, l/n_0) = 1. \tag{2.15}$$

2.3 The matrix relations for a single tetrahedron

Using the properties of interpolation functions described above it is clear that an approximation of degree n_0,

$$U = U_{ijkl} \alpha_{ijkl}(\zeta_1, \zeta_2, \zeta_3, \zeta_4), \tag{2.16}$$

can be made for a potential function U within a tetrahedron. Here, i, j, k and l are positive integers or zero whose sum is n_0 and U_{ijkl} represents the as yet undetermined set of values of the function U at $N = (n_0+1)(n_0+2)(n_0+3)/6$ regularly spaced interpolation points within or on the faces of the tetrahedron. Using arguments similar to those put forward in the two-dimensional case of Chapter 3, continuity of U is assured across the boundaries of abutting tetrahedra in an assembly of finite elements of like order N. A source term g as in Eq. (2.03) may also be approximated in terms of the interpolation functions by

$$g = \sum_{ijkl} g_{ijkl} \alpha_{ijkl}. \tag{2.17}$$

The approximations (2.16) and (2.17) can be substituted into the expression (2.04) defining the functional F associated with the inhomogeneous Helmholtz equation. Then the requirement for stationarity,

$$\frac{\partial F}{\partial U_m} = 0, \tag{2.18}$$

for the N values of the multi-index $m = ijkl$, is applied. The constraint of Eq. (2.18) yields a set of values $U_m = U_{ijkl}$ which may reasonably be assumed to represent the best fit of the polynomial functions (2.16) and (2.17) to the Helmholtz equation, Eq. (2.03). It will now be shown that the result of this constraint is a matrix equation

$$\mathbf{SU} - k^2 \mathbf{TU} + \mathbf{TG} = 0, \tag{2.19}$$

in which U_m and g_m are represented as vectors $\mathbf{U}$ and $\mathbf{G}$. The result is a three-dimensional extension of Eq. (4.10) of Chapter 3. For, Eqs. (2.16) and (2.17) may be written in single-index form

$$U = \sum_{m=1}^{N} U_m \alpha_m, \tag{2.20}$$

$$g = \sum_{m=1}^{N} g_m \alpha_m. \tag{2.21}$$

Then

$$\nabla U = \sum_{m=1}^{N} U_m \nabla \alpha_m \tag{2.22}$$

and

$$F(\mathbf{U}) = \tfrac{1}{2} p \sum_{m=1}^{N} \sum_{n=1}^{N} U_m U_n \int_{\Omega} (\nabla \alpha_m) \cdot (\nabla \alpha_n) \, d\Omega$$

$$- \tfrac{1}{2} k^2 \sum_{m=1}^{N} \sum_{n=1}^{N} U_m U_n \int_{\Omega} \alpha_m \alpha_n \, d\Omega$$

$$+ \sum_{m=1}^{N} \sum_{n=1}^{N} U_m \int_{\Omega} \alpha_m g_n \alpha_n \, d\Omega, \tag{2.23}$$

where it is assumed that the material property p remains constant within the tetrahedron. Written in matrix form Eq. (2.33) is

$$F(\mathbf{U}) = \tfrac{1}{2}\mathbf{U}^T\mathbf{S}\mathbf{U} - \tfrac{1}{2}k^2\mathbf{U}^T\mathbf{T}\mathbf{U} + \mathbf{U}^T\mathbf{T}\mathbf{G}, \qquad (2.24)$$

where

$$S_{mn} = p \int_\Omega (\nabla\alpha_m)\cdot(\nabla\alpha_n)\,d\Omega$$

$$= p \int_\Omega \left(\frac{\partial\alpha_m}{\partial x}\frac{\partial\alpha_n}{\partial x} + \frac{\partial\alpha_m}{\partial y}\frac{\partial\alpha_n}{\partial y} + \frac{\partial\alpha_m}{\partial z}\frac{\partial\alpha_n}{\partial z} \right) d\Omega \qquad (2.25)$$

and

$$T_{mn} = \int_\Omega \alpha_m\alpha_n\,d\Omega. \qquad (2.26)$$

It is now seen that applying the condition for stationarity, Eqs. (2.18)–(2.23) gives the matrix equation, Eq. (2.19). It remains to evaluate the matrices $\mathbf{S}$ and $\mathbf{T}$, three-dimensional generalisations of the matrices so-designated in Chapter 3.

The integrations of Eqs. (2.23) and (2.24) are to be taken over the tetrahedron of volume Ω with respect to the three-dimensional spatial coordinates, say Cartesians (x, y, z), so that $d\Omega = dx\,dy\,dz$. The formal procedure for transforming to an integral in the homogeneous coordinate space of $\zeta_1, \zeta_2, \zeta_3$ and $\zeta_4 = 1 - \zeta_1 - \zeta_2 - \zeta_3$, requires $dx\,dy\,dz$ to be replaced by

$$d\zeta_1\,d\zeta_2\,d\zeta_3 \left/ \frac{\partial(\zeta_1, \zeta_2, \zeta_3)}{\partial(x, y, z)} \right. ,$$

just as in the two-dimensional case treated in the Appendix. Here

$$\frac{\partial(\zeta_1, \zeta_2, \zeta_3)}{\partial(x, y, z)} = \begin{vmatrix} \dfrac{\partial\zeta_1}{\partial x} & \dfrac{\partial\zeta_1}{\partial y} & \dfrac{\partial\zeta_1}{\partial z} \\[2mm] \dfrac{\partial\zeta_2}{\partial x} & \dfrac{\partial\zeta_2}{\partial y} & \dfrac{\partial\zeta_2}{\partial z} \\[2mm] \dfrac{\partial\zeta_3}{\partial x} & \dfrac{\partial\zeta_3}{\partial y} & \dfrac{\partial\zeta_3}{\partial z} \end{vmatrix} \qquad (2.27)$$

is the Jacobian determinant for the coordinate transformation Eq. (2.12). The choice of ζ_4 as the homogeneous variable to be eliminated using Eq. (2.08) is entirely arbitrary. Any other of the set $\zeta_1, \zeta_2, \zeta_3, \zeta_4$ could have been selected. It may be shown that the magnitude of the Jacobian (2.26) is $1/(3!\,V)$ so that $d\Omega = dx\,dy\,dz$ must be replaced by $3!\,V\,d\zeta_1\,d\zeta_2\,d\zeta_3$ in order to perform the volume integrals of Eqs. (2.24) and (2.25) with respect to the homogeneous coordinates $\zeta_1, \zeta_2, \zeta_3$ and

$\zeta_4 = 1 - \zeta_1 - \zeta_2 - \zeta_3$. Thus, with the aid of Eq. (2.12) and the chain rule of differentiation, Eq. (2.23) may be rewritten

$$S_{mn} = \sum_{i=1}^{4} \sum_{j=1}^{4} pK_{ij} \int 6 \frac{\partial \alpha_m}{\partial \zeta_i} \frac{\partial \alpha_n}{\partial \zeta_j} \, d\zeta_1 \, d\zeta_2 \, d\zeta_3, \qquad (2.28)$$

where

$$K_{ij} = \frac{(b_i b_j + c_i c_j + d_i d_j)}{36 V} \qquad (2.29)$$

It is understood that wherever ζ_4 appears explicitly in the polynomial expressions of Eq. (2.28) it is to be replaced by $1 - \zeta_1 - \zeta_2 - \zeta_3$.

A notation

$$Q_{mn}^{ij} = -6 \int_{\Omega} \left(\frac{\partial \alpha_m}{\partial \zeta_i} - \frac{\partial \alpha_m}{\partial \zeta_j} \right) \left(\frac{\partial \alpha_n}{\partial \zeta_i} - \frac{\partial \alpha_n}{\partial \zeta_j} \right) d\zeta_1 \, d\zeta_2 \, d\zeta_3, \qquad (2.30)$$

extending into three dimensions the two-dimensional formalism adopted in Chapter 3, is introduced.

It is found that Eq. (2.28) can be written as

$$S_{mn} = \sum_{i=1}^{3} \sum_{j=i+1}^{4} pK_{ij} Q_{mn}^{ij}, \qquad (2.31)$$

where Q_{mn}^{ij} is a purely numeric array which may be calculated once and for all for any order n_0 of the polynomial interpolation approximation, whilst the array of K_{ij} contains all the specific information associated with the geometric shape, size and position of the tetrahedron. The summation of Eq. (2.31) involves just six index pairs (i, j), $i \neq j$, corresponding to the six tetrahedron edges, the terms K_{ii} being absent. Equation (2.31) may be verified by first observing that the linear interdependence of the four homogeneous coordinates, $\sum \zeta_i = 1$ requires that the coefficients b, c and d in Eq. (2.12) and in Eq. (2.02) each should sum to zero,

$$\sum_{i=1}^{4} b_i = \sum_{i=1}^{4} c_i = \sum_{i=1}^{4} d_i = 0. \qquad (2.32)$$

Thus it is clear that the diagonal elements of the array K_{ij}, $i = j$, need not be specified independently but may be calculated from the off-diagonal terms K_{ij}, $i \neq j$, and that

$$K_{11} = -(K_{12} + K_{13} + K_{14})$$
$$K_{22} = -(K_{21} + K_{23} + K_{24}), \qquad (2.33)$$

and so forth. Further observing that the array is symmetric, $K_{ij} = K_{ji}$, Eq. (2.28) may be written out in full, the diagonal terms K_{ii} being replaced by Eq. (2.33) and lower triangle terms such as K_{21} being substituted by K_{12}. It is then evident by arranging the grouping of terms, that the upper triangle, off-diagonal summation of Eq. (2.31) amounts to exactly the same as Eq. (2.28) so written out.

2.4. *Evaluation of the matrices* **T** *and* **Q**

Equations (2.30) and (2.26), together with the definition of the interpolation polynomials $\alpha_m(\zeta_1, \zeta_2, \zeta_3, \zeta_4)$ given by Eqs. (2.13) and (2.14), represent a sequence of algebraic operations. If carried out these allow the matrix coefficients T_{mn} and Q_{mn}^{ij} to be established for m and n each in the set $\{1, 2, 3, \ldots, N\}$, $N = (n_0 + 1)(n_0 + 2)(n_0 + 3)/6$, where n_0 is the order of the polynomial approximation, and for i and j each in the set $\{1, 2, 3, 4\}$, $i \neq j$. In carrying out such operations, integrals of the form

$$I(i, j, k, l) = \int_\Omega \zeta_1^i \zeta_2^j \zeta_3^k \zeta_4^l \, d\zeta_1 \, d\zeta_2 \, d\zeta_3 \tag{2.34}$$

are required. More explicitly

$$I(i, j, k, l)$$
$$= \int_0^1 d\zeta_1 \int_0^{1-\zeta_1} d\zeta_2 \int_0^{1-\zeta_1-\zeta_2} \zeta_1^i \zeta_2^j \zeta_3^k (1 - \zeta_1 - \zeta_2 - \zeta_3)^l \, d\zeta_3. \tag{2.35}$$

Straightforward integration by parts, successively reducing an index in Eq. (2.34) by unity each time, gives the result

$$I(i, j, k, l) = \frac{i! \, j! \, k! \, l! \, 3!}{(i + j + k + l + 3)!}. \tag{2.36}$$

Evidently, using Eq. (2.26), the array T_{mn} is obtained from a 'once and for all' purely numeric matrix multiplied by the tetrahedron volume:

$$T_{mn} = V \int_\Omega \alpha_m \alpha_n \, d\zeta_1 \, d\zeta_2 \, d\zeta_3. \tag{2.37}$$

In calculating the numeric T_{mn} normalised with respect to V it is evident that the matrix **T** is symmetric, $T_{mn} = T_{nm}$. However, more symmetries exist, resulting in further computation economy, because the normalised **T** is entirely unbiased with respect to the particular labelling of the homogeneous coordinates ζ_i, $i = 1, 2, 3, 4$. These symmetries are demonstrated by taking as an example the second-order tetrahedron shown in Fig. 6.2.

The tetrahedron vertices can be identified by the index i (or j) = 1, 2, 3, 4. A tetrahedron edge is represented by one of the six pairs (i, j). The $N = 10$ nodal points involved are shown with their multi-index labelling $(pqrs)$, $p + q + r + s = 2$. Also shown is an arbitrarily chosen single-index scheme defining m (or n). The fact that a systematic single indexing is followed is immaterial, although the 'code' used does need to be known when employing matrices presented in the single-index form. The single-index scheme here and its logical extensions to higher orders is in fact the one adopted by Silvester (1972). Examination of Fig. 6.2 reveals that only T_{11}, T_{22}, T_{12}, T_{15}, T_{16}, T_{23} and T_{29} need to be evaluated separately in this second-order case. All the rest of the 100 elements involved will assume one or other of the seven numerical values above, because these remaining pairs of indices could always be identified with one of the given seven by viewing the tetrahedron differently and relabelling the multi-indexes $(pqrs)$ with an appropriate permutation. Turning to the matrices $\mathbf{Q}^{ij}$, it may be observed that the coefficients Q_{mn}^{ij} for each of the six pairs (i, j), $i > j$, which represent tetrahedron edges, are required in order to set up the array S_{mn} from Eq. (2.31). The coefficients Q_{mn}^{ij} are independent of the shape and size of the tetrahedron whereas they are evidently determined by the nature of the juxtaposition of the edge (i, j) with respect to the nodal points m and n. The actual labelling of the edges and points should have no bearing upon the resulting numbers and it follows that each array $\mathbf{Q}^{ij}$ should be obtainable from permutations of one set of numbers, $\mathbf{Q}^{12}$ say. The procedure for obtaining all of the $\mathbf{Q}^{ij}$-matrices from $\mathbf{Q}^{12}$ is best explained with reference to a specific example. The case $n_0 = 2$, $N = 10$ with the single-index numbering already employed is chosen again and illustrated in Fig. 6.3.

Suppose that the set of numbers Q_{mn}^{12} has been worked out. It is clear by comparing the diagrams (a) and (b) of Fig. 6.3 that, for instance, the equation

$$Q_{46}^{14} = Q_{29}^{12} \tag{2.38}$$

holds, since the numbers in the $\mathbf{Q}$-matrix are determined only by the relative positions of points m, n and the edge (i, j) and not by the particular labels assigned to these. In the case here, the same number Q in question represents the edge shown heavily lined, combined with

Fig. 6.2. Tetrahedron showing nodal numbering schemes for a second-order polynomial approximation.

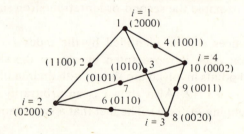

the two circled nodal points, however these may be identified. It may be confirmed that Eq. (2.38) is a particular equation taken from the complete matrix element permutation

$$\mathbf{Q}^{14} = \mathbf{R}_1 \mathbf{Q}^{12} \mathbf{R}_1^{\mathrm{T}}, \qquad (2.39)$$

where $\mathbf{R}_1$ is a rotation matrix

$$\mathbf{R}_1 = \begin{bmatrix} 1 & & & & & & & & & \\ & 1 & & & & & & & & \\ & & 1 & & & & & & & \\ & & & 1 & & & & & & \\ & & & & & & 1 & & & \\ & & & & & & & 1 & & \\ & & & & & 1 & & & & \\ & & & & & & & & 1 & \\ & & & & 1 & & & & & \\ & & & 1 & & & & & & \end{bmatrix} \qquad (2.40)$$

and $\mathbf{R}_1^{\mathrm{T}}$ is its transpose. The notation $\mathbf{R}_1$ refers to a permutation of vertex indices corresponding to a right-handed rotation about the ζ_1-axis, that is to say the vertex 1 remains so-labelled whilst vertices 2, 3 and 4 are relabelled 4, 2 and 3 respectively. On this scheme the $\mathbf{R}_2$, $\mathbf{R}_3$ and $\mathbf{R}_4$-permutations relabel the vertices $1, 2, 3, 4$ as $3, 2, 4, 1$, $4, 1, 3, 2$ and $2, 3, 1, 4$ respectively. It may be confirmed by extending the geometric argument embodied in Fig. 6.3 and noting that $\mathbf{Q}^{ij} = \mathbf{Q}^{ji}$, that the five other matrices required in addition to $\mathbf{Q}^{12}$ are

$$\begin{aligned} \mathbf{Q}^{14} &= \mathbf{R}_1 \mathbf{Q}^{12} \mathbf{R}_1^{\mathrm{T}} \\ \mathbf{Q}^{13} &= \mathbf{R}_1 \mathbf{Q}^{14} \mathbf{R}_1^{\mathrm{T}} \\ \mathbf{Q}^{23} &= \mathbf{R}_2 \mathbf{Q}^{12} \mathbf{R}_2^{\mathrm{T}} \\ \mathbf{Q}^{24} &= \mathbf{R}_2 \mathbf{Q}^{23} \mathbf{R}_2^{\mathrm{T}} \\ \mathbf{Q}^{34} &= \mathbf{R}_1 \mathbf{Q}^{24} \mathbf{R}_1^{\mathrm{T}}. \end{aligned} \qquad (2.41)$$

Of course, all 100 elements of $\mathbf{Q}^{12}$ do not have to be computed from

Fig. 6.3. (*a*) Second-order tetrahedron. (*b*) The same viewed after a rotation about the ζ_1-axis.

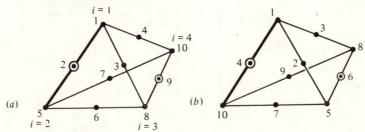

first principles. Because of the symmetries of the type discussed with respect to the matrix $\mathbf{T}$, relatively few of the numbers Q_{mn}^{12} need to be calculated separately. The unnecessary independent evaluations for the $n_0 = 2$ case here are easily picked out by inspection from Fig. 6.3(a); for instance, clearly, $Q_{13}^{12} = Q_{14}^{12} = Q_{56}^{12} = Q_{57}^{12}$. The matrices $\mathbf{T}$ (normalised to volume V) and $\mathbf{Q}^{12}$ corresponding to the single-index scheme illustrated in Figs. 6.2 and 6.3 are tabulated in Fig. 6.4 for $n_0 = 1$ and 2. Since the matrices are both symmetric, only the upper triangle is given for $\mathbf{T}$ and the lower for $\mathbf{Q}^{12}$.

The tabulated numbers are numerators of integer quotients; the common denominators are given separately. The somewhat unwieldy matrices for $n_0 = 3$ have also been worked out and are given by Silvester (1972). The operations involved in generating the matrices of order $n_0 = 2$ and higher are complicated and must be done by computer. Integer arithmetic is used for preference in order to eliminate the roundoff uncertainty in presenting fractional numbers.

2.5. Assembly of global matrix equations

Having set up S_{mn} and T_{mn} for a single tetrahedron in an ensemble of elements constituting the three-dimensional space of some problem, there is no fundamental difficulty in obtaining the 'global' matrix representing the connected set of elements. The procedure follows exactly that used in two dimensions (Chapter 1, Section 4 and Chapter 3, Section 5). A matrix $\mathbf{C}$ is set up to express the constraints imposed upon the nodal potential values $\mathbf{U}_{\text{dis}}$ associated with each tetrahedron numbered separately. Thus if

$$\mathbf{U}_{\text{dis}} = \mathbf{C}\mathbf{U}_{\text{con}}, \tag{2.42a}$$

$$\mathbf{G}_{\text{dis}} = \mathbf{C}\mathbf{G}_{\text{con}}, \tag{2.42b}$$

Fig. 6.4. The matrices T normalised to volume (upper triangles) and Q^{12} (lower triangles) for orders $n_0 = 1$ and 2.

$n_0 = 1$

```
      2   1   1   1
 -1       2   1   1
  1  -1       2   1
  0   0   0       2
  0   0   0   0
```

Denominators: $D(Q) = 1$; $D(T) = 20$

$n_0 = 2$

```
        6  -4  -4  -4   1  -6  -6   1  -6   1
-12        32  16  16  -4  16  16  -6   8  -6
 16 -32        32  16  -6  16   8  -4  16  -6
  4   0 -32        32  -6   8  16  -6  16  -4
  4   0 -16 -32         6  -4  -4   1  -6   1
 -4  16  -4  -4 -12        32  16  -4  16  -6
 -4   0  32  16   4 -32        32  -6  16  -4
 -4   0  16  32   4 -16 -32         6  -4   1
  0   0   0   0   0   0   0   0        32  -4
  0   0   0   0   0   0   0   0   0         6
```

Denominators: $D(Q) = 20$; $D(T) = 420$

then with
$$\mathbf{S} = \mathbf{C}^{T}\mathbf{S}_{dis}\mathbf{C} \tag{2.43}$$
and
$$\mathbf{T} = \mathbf{C}^{T}\mathbf{T}_{dis}\mathbf{C}, \tag{2.44}$$
an equation corresponding to Eq. (2.19),
$$\mathbf{S}\mathbf{U}_{con} - k^{2}\mathbf{T}\mathbf{U}_{con} + \mathbf{T}\mathbf{G}_{con} = 0, \tag{2.45}$$
must be satisfied. This connection procedure, it is recalled, arises essentially from the additive nature of the contributions $F(\mathbf{U})$ of separate elements, defined by Eq. (2.24), to the global functional for the whole of the problem-space. Having obtained global matrices $\mathbf{S}$ and $\mathbf{T}$ for some particular problem the solutions for the potential $\mathbf{U}_f$ at free nodes and for the eigenvalues k (if any) follow exactly on the lines adopted in two dimensions (Chapter 1, Section 4 and Chapter 3, Section 6). Once again, the homogeneous Neumann boundary condition, $\partial U/\partial n = 0$, arises naturally if the potentials to be calculated at Neumann boundary points are left unconstrained, just as if they had been interior nodes.

2.6   Matrix equations for a rectangular prism

In this section the $\mathbf{S}$- and $\mathbf{T}$-matrices applying to the eight vertex nodes of a rectangular prism having sides a, b and c are established. The example here is chosen to illustrate the practical steps which need to be taken in order to apply the preceding analysis to a real problem. It is shown in Section 2.7 that results corresponding to the well-known simple physics and analysis of a rectangular prism acoustic cavity are achieved. The problem considered is one of the rare instances where the calcuation can proceed without the use of a computer. Nevertheless, it exhibits many of the features which programming for more complicated geometries reveal. In order to obtain the matrix relations the prism is considered to be made up from five connected, first-order tetrahedra as shown in Fig. 6.5. Inspection of Fig. 6.5 reveals that the prism has been dissected such that four of the five tetrahedra are congruent in that any one of them could, with a translation and a rotation, be made to fit over any of the others, vertex-for-vertex. A nodal numbering scheme consistent with this congruency has been adopted for elements 1–4, which are each seen to have volume $abc/6$. The fifth element, interior to the prism apart from its edges, is different from the rest and has volume $abc/3$. A connection matrix $\mathbf{C}$ is defined by Eq. (2.42), $\mathbf{U}_{dis} = \mathbf{C}\mathbf{U}_{con}$, where $\mathbf{U}_{dis}$ is the vector of twenty U-values for the five disconnected elements and $\mathbf{U}_{con}$ the vector corresponding to the eight prism corner-nodes. By inspection

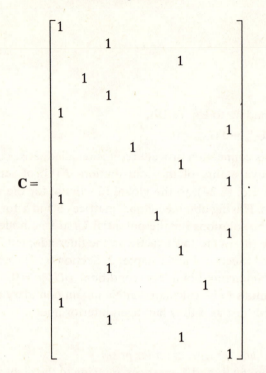

$$C = [\ \ldots\ ] . \tag{2.46}$$

Fig. 6.5. (*a*) Complete prism, volume *abc*. (*b*)–(*e*) Elements 1, 2, 3 and 4 are shown respectively. They are congruent and each has volume *abc*/6. (*f*) Element 5. It has volume *abc*/3. The numbering in (*a*) refers to the connected set of tetrahedra, whereas that in (*b*)–(*f*) follows the unconnected element scheme.

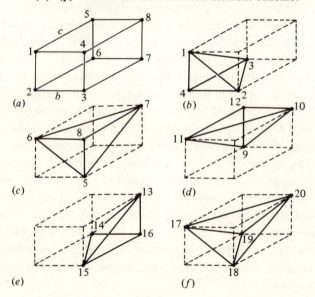

Using the $n_0 = 1$ normalised **T**-matrix given in Fig. 6.4 it is seen immediately that

$$\mathbf{T}_{dis} = \frac{abc}{120}
\begin{bmatrix}
2 & 1 & 1 & 1 & & & & & & & & & & & & \\
1 & 2 & 1 & 1 & & & & & & & & & & & & \\
1 & 1 & 2 & 1 & & & & & & & & & & & & \\
1 & 1 & 1 & 2 & & & & & & & & & & & & \\
& & & & 2 & 1 & 1 & 1 & & & & & & & & \\
& & & & 1 & 2 & 1 & 1 & & & & & & & & \\
& & & & 1 & 1 & 2 & 1 & & & & & & & & \\
& & & & 1 & 1 & 1 & 2 & & & & & & & & \\
& & & & & & & & 2 & 1 & 1 & 1 & & & & \\
& & & & & & & & 1 & 2 & 1 & 1 & & & & \\
& & & & & & & & 1 & 1 & 2 & 1 & & & & \\
& & & & & & & & 1 & 1 & 1 & 2 & & & & \\
& & & & & & & & & & & & 2 & 1 & 1 & 1 \\
& & & & & & & & & & & & 1 & 2 & 1 & 1 \\
& & & & & & & & & & & & 1 & 1 & 2 & 1 \\
& & & & & & & & & & & & 1 & 1 & 1 & 2 \\
& & & & & & & & & & & & & & & & 4 & 2 & 2 & 2 \\
& & & & & & & & & & & & & & & & 2 & 4 & 2 & 2 \\
& & & & & & & & & & & & & & & & 2 & 2 & 4 & 2 \\
& & & & & & & & & & & & & & & & 2 & 2 & 2 & 4
\end{bmatrix} .$$

$$(2.47)$$

Further, using the relationship

$$\mathbf{T}_{con} = \mathbf{C}^{\mathrm{T}} \mathbf{T}_{dis} \mathbf{C}, \tag{2.48}$$

the reader may readily verify using matrix expressions (2.46) and (2.47) that

$$\mathbf{T}_{con} = \frac{abc}{120}
\begin{bmatrix}
10 & 1 & 4 & 1 & 1 & 4 & 0 & 4 \\
1 & 2 & 1 & 0 & 0 & 1 & 0 & 0 \\
4 & 1 & 10 & 1 & 0 & 4 & 1 & 4 \\
1 & 0 & 1 & 2 & 0 & 0 & 0 & 1 \\
1 & 0 & 0 & 0 & 2 & 1 & 0 & 1 \\
4 & 1 & 4 & 0 & 1 & 10 & 1 & 4 \\
0 & 0 & 1 & 0 & 0 & 1 & 2 & 1 \\
4 & 0 & 4 & 1 & 1 & 4 & 1 & 10
\end{bmatrix} . \tag{2.49}$$

The determination of $\mathbf{S}_{con}$ is similarly straightforward but a little more lengthy. Attention is first focused upon element 1, redrawn in Fig. 6.6

with a set of Cartesian axes superimposed. Clearly homogeneous co-ordinates here are expressed by

$$\zeta_1 = x/a, \qquad \zeta_2 = y/b, \qquad \zeta_3 = z/c, \qquad \zeta_4 = 1 - x/a - y/b - z/c. \quad (2.50)$$

Thus since, in general,

$$\zeta_i = \frac{a_i + b_i x + c_i y + d_i z}{6V} \qquad (2.51)$$

(see Eq. (2.12)) where V is the element volume, here $abc/6$, it follows that

$$
\begin{aligned}
b_1 &= bc, & c_1 &= 0, & d_1 &= 0 \\
b_2 &= 0, & c_2 &= ac, & d_2 &= 0 \\
b_3 &= 0, & c_3 &= 0, & d_3 &= ab \\
b_4 &= -bc, & c_4 &= -ac, & d_4 &= -ab.
\end{aligned}
$$

From Eq. (2.29), substituting in the element volume appropriate here,

$$K_{ij} = \frac{b_i b_j + c_i c_j + d_i d_j}{6abc}, \qquad (2.52)$$

so that

$$K_{14} = -bc/6a, \qquad K_{24} = -ac/6b, \qquad K_{34} = -ab/6c$$
$$K_{12} = K_{13} = K_{23} = 0.$$

From Fig. 6.4 it is seen that

$$
\mathbf{Q}^{12} = \begin{bmatrix} -1 & 1 & 0 & 0 \\ 1 & -1 & 0 & 0 \\ 0 & 0 & 0 & 0 \\ 0 & 0 & 0 & 0 \end{bmatrix}. \qquad (2.53)
$$

In order to evaluate the five other $\mathbf{Q}$-matrices the rotation matrices $\mathbf{R}$ of Section 2.4 could be written down and $\mathbf{Q}^{12}$ transformed appropriately.

Fig. 6.6. Element 1 with superimposed Cartesian axes.

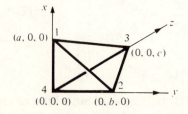

However, in this simple case it is seen that $\mathbf{Q}^{ij}$ is described by saying that $Q_{ii}^{ij} = Q_{jj}^{ij} = -1$ and $Q_{ij}^{ij} = Q_{ji}^{ij} = 1$, with all other terms zero. Thus the S-matrix for element 1 is, from Eq. (2.31)

$$\mathbf{S}_1 = K_{14}\mathbf{Q}^{14} + K_{24}\mathbf{Q}^{24} + K_{34}\mathbf{Q}^{34}$$

$$= -\frac{bc}{6a}\begin{bmatrix} -1 & 0 & 0 & 1 \\ 0 & 0 & 0 & 0 \\ 0 & 0 & 0 & 0 \\ 1 & 0 & 0 & 1 \end{bmatrix} - \frac{ac}{6b}\begin{bmatrix} 0 & 0 & 0 & 0 \\ 0 & -1 & 0 & 1 \\ 0 & 0 & 0 & 0 \\ 0 & 1 & 0 & -1 \end{bmatrix}$$

$$- \frac{ab}{6c}\begin{bmatrix} 0 & 0 & 0 & 0 \\ 0 & 0 & 0 & 0 \\ 0 & 0 & -1 & 1 \\ 0 & 0 & 1 & -1 \end{bmatrix}, \tag{2.54}$$

$$= \frac{abc}{6}\begin{bmatrix} \dfrac{1}{a^2} & 0 & 0 & -\dfrac{1}{a^2} \\ 0 & \dfrac{1}{b^2} & 0 & -\dfrac{1}{b^2} \\ 0 & 0 & \dfrac{1}{c^2} & -\dfrac{1}{c^2} \\ -\dfrac{1}{a^2} & -\dfrac{1}{b^2} & -\dfrac{1}{c^2} & \left(\dfrac{1}{a^2}+\dfrac{1}{b^2}+\dfrac{1}{c^2}\right) \end{bmatrix}. \tag{2.55}$$

Equation (2.55) is written for conciseness

$$\mathbf{S}_1 = \frac{abc}{6}\begin{bmatrix} X & 0 & 0 & -X \\ 0 & Y & 0 & -Y \\ 0 & 0 & Z & -Z \\ -X & -Y & -Z & A \end{bmatrix}, \tag{2.56}$$

where $X = 1/a^2$, $Y = 1/b^2$, $Z = 1/c^2$ and $A = X + Y + Z$. The S-matrices S_2, S_3 and S_4 for elements 2–4 are identical.

Attention is now focused upon element 5, redrawn in Fig. 6.7 with Cartesian axes superimposed. In this case Eqs. (2.07), (2.09) and (2.10) give the homogeneous coordinate ζ_1 as

$$\zeta_1 = \frac{1}{6}\begin{vmatrix} 1 & x & y & z \\ 1 & 0 & b & 0 \\ 1 & 0 & 0 & c \\ 1 & a & b & c \end{vmatrix} \div \left(\frac{abc}{3}\right), \tag{2.57}$$

$$= (1 + x/a - y/b - z/c)/2. \tag{2.58}$$

Similarly,

$$\zeta_2 = (1 - x/a + y/b - z/c)/2, \tag{2.59}$$
$$\zeta_3 = (1 - x/a - y/b + z/c)/2, \tag{2.60}$$
$$\zeta_4 = (-1 + x/a + y/b + z/c)/2. \tag{2.61}$$

Application of Eqs. (2.12) and (2.29), remembering that the element volume here is $V = abc/3$ and using the notation of Eq. (2.56), gives

$$\left.\begin{aligned}
K_{12} &= \frac{abc}{6}\left(-\frac{A}{2} + Z\right) \\[6pt]
K_{13} &= \frac{abc}{6}\left(-\frac{A}{2} + Y\right) \\[6pt]
K_{14} &= \frac{abc}{6}\left(-\frac{A}{2} + X\right) \\[6pt]
K_{23} &= \frac{abc}{6}\left(-\frac{A}{2} + X\right) \\[6pt]
K_{24} &= \frac{abc}{6}\left(-\frac{A}{2} + Y\right) \\[6pt]
K_{34} &= \frac{abc}{6}\left(-\frac{A}{2} + Z\right).
\end{aligned}\right\} \tag{2.62}$$

Using Eq. (2.31) and noting that $\mathbf{Q}^{ij}$ is always a simple permutation of the matrix (2.53) for $\mathbf{Q}^{12}$, it is readily found that for element 5

$$\mathbf{S}_5 = \frac{abc}{6}\begin{bmatrix}
\dfrac{A}{2}, & -\dfrac{A}{2} + Z, & -\dfrac{A}{2} + Y, & -\dfrac{A}{2} + X \\[8pt]
-\dfrac{A}{2} + Z, & \dfrac{A}{2}, & -\dfrac{A}{2} + X, & -\dfrac{A}{2} + Y \\[8pt]
-\dfrac{A}{2} + Y, & -\dfrac{A}{2} + X, & \dfrac{A}{2}, & -\dfrac{A}{2} + Z \\[8pt]
-\dfrac{A}{2} + X, & -\dfrac{A}{2} + Y, & -\dfrac{A}{2} + Z, & \dfrac{A}{2}
\end{bmatrix}. \tag{2.63}$$

Fig. 6.7. Element 5 with superimposed Cartesian axes.

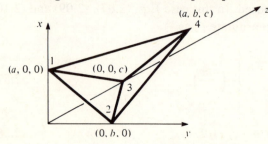

Thus the disconnected form of the matrix **S** is written down as

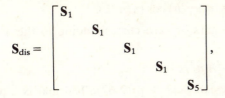

$$\mathbf{S}_{dis} = \begin{bmatrix} \mathbf{S}_1 & & & & \\ & \mathbf{S}_1 & & & \\ & & \mathbf{S}_1 & & \\ & & & \mathbf{S}_1 & \\ & & & & \mathbf{S}_5 \end{bmatrix},$$

where $\mathbf{S}_1$ and $\mathbf{S}_5$ are 4×4 matrices defined by Eqs. (2.56) and (2.63). The connected matrix required, $\mathbf{S}_{con} = \mathbf{C}^T \mathbf{S}_{dis} \mathbf{C}$, may now be written down, observing that **C** remains as defined by Eq. (2.46). The algebra involved is not prodigious although it does require a cool head! The result is found to be

$$\mathbf{S}_{con} = \frac{abc}{6} \begin{bmatrix} 3A/2 & -X & Z-A/2 & -Y & -Z & Y-A/2 & 0 & X-A/2 \\ -X & A & -Y & 0 & 0 & -Z & 0 & 0 \\ Z-A/2 & -Y & 3A/2 & -X & 0 & X-A/2 & -Z & Y-A/2 \\ -Y & 0 & -X & A & 0 & 0 & 0 & -Z \\ -Z & 0 & 0 & 0 & A & -X & 0 & -Y \\ Y-A/2 & -Z & X-A/2 & 0 & -X & 3A/2 & -Y & Z-A/2 \\ 0 & 0 & -Z & 0 & 0 & -Y & S & -X \\ X-A/2 & 0 & Y-A/2 & -Z & -Y & Z-A/2 & -X & 3A/2 \end{bmatrix}. \quad (2.65)$$

Having obtained the **S**- and **T**-matrices for a basic 'brick' of side a, b, c one is now in a position to build up three-dimensional shapes from the bricks, using further simple connecting operations. Thus three-dimensional problems involving the single-variable inhomogeneous Helmholtz equation may be solved by matrix techniques.

2.7 *Application to a cubic acoustic resonator*

It is interesting to consider the very rudimentary result which can be obtained for a cubic acoustic resonator by considering just one 'brick'. The partial differential equation governing the problem is shown as Eq. (8.11) in Chapter 2 to be

$$\nabla^2 p + k^2 p = 0, \quad (2.66)$$

where the excess pressure p varying at frequency ω is subject to the Neumann boundary condition $\partial p / \partial n = 0$ at the rigid-wall boundaries. The wavenumber k is related to resonant frequency ω and speed of sound v by

$$k = \omega / v. \quad (2.67)$$

The reader may easily check that an analytical solution for the rectangular prism, normalised to unit amplitude, is

$$p = \cos(l\pi x/a)\cos(m\pi y/b)\cos(n\pi z/c), \tag{2.68}$$

where l, m and n are integers or zero corresponding to the acoustic cavity modes and where

$$k^2 = (l^2/a^2 + m^2/b^2 + n^2/c^2)\pi^2 \tag{2.69}$$

sets the resonant frequency through Eq. (2.67). The matrix equation equivalent of Eq. (2.66) is

$$\mathbf{S}\mathbf{U} - k^2\mathbf{T}\mathbf{U} = 0, \tag{2.70}$$

where, if $\mathbf{U}$ is associated with the vector of values of p at the eight nodes of a rectangular prism then, $\mathbf{S}$ and $\mathbf{T}$ are the matrices of Eqs. (2.49) and (2.65) respectively. Although each of the eight nodes is a boundary point, they are all unconstrained because the Neumann boundary condition applies. Thus the eigenvalues k corresponding to acoustic modes derive from the vanishing of the determinant

$$\Delta = |\mathbf{S} - k^2\mathbf{T}|. \tag{2.71}$$

For a cube of side a, writing $\lambda = k^2a^2/20$ and putting $X = Y = Z = 1/a^2$, $A = 3/a^2$, it is found that

$$\Delta = \frac{a^3}{6}\begin{vmatrix} \alpha & \gamma & \delta & \gamma & \gamma & \delta & 0 & \delta \\ \gamma & \beta & \gamma & 0 & 0 & \gamma & 0 & 0 \\ \delta & \gamma & \alpha & \gamma & 0 & \delta & \gamma & \delta \\ \gamma & 0 & \gamma & \beta & 0 & 0 & Q & \gamma \\ \gamma & 0 & 0 & 0 & \beta & \gamma & 0 & \gamma \\ \delta & \gamma & \delta & 0 & \gamma & \alpha & \gamma & \delta \\ 0 & 0 & \gamma & 0 & 0 & \gamma & \beta & \gamma \\ \delta & 0 & \delta & \gamma & \gamma & \delta & \gamma & \alpha \end{vmatrix}, \tag{2.72}$$

where $\alpha = 9/2 - 10\lambda$, $\beta = 3 - 2\lambda$, $\gamma = -(1+\lambda)$ and $\delta = 1/2 - 4\lambda$. In this example, clearly, the condition for Δ to vanish will involve an eighth-order algebraic equation in λ, which will have eight roots giving estimates of k for the first eight acoustic modes. Normally, eigenvalue-solving computer routines have to be employed to obtain the roots. Here, however, because of the extreme simplicity of the problem the equations can be written down and solved by hand. To do this first it is noted that the matrix $\mathbf{S}$ always has zero row (and column) sums, so that columns of the determinant Δ, Eq. (2.72), sum alternately to -27λ and -5λ. With further row and column manipulations it is found that

$$\Delta = -5\lambda[\gamma^2 + (\delta - \alpha)\beta]^3(3\gamma - 5\beta)a^3/6$$
$$= -5\lambda(-11\lambda^2 + 28\lambda - 11)^3(7\lambda - 18)a^3/6. \tag{2.73}$$

Equation (2.73) yields two distinct single roots and a pair of triple roots. Table 6.1 shows that this is just what would be expected in the light of the analytical solution Eq. (2.68). The trivial solution $p = $ constant appears once, then the triply-degenerate fundamental mode is predicted with very reasonable accuracy. Thereafter, the mode identification is preserved, that is to say, a further triply-degenerate mode and then a single unique mode are predicted as would be expected, but the accuracy of the k-values falls off. Of course, no more than eight from the infinite set of modes can be predicted by the simple model here which has only eight degrees of freedom.

3. Three-dimensional problems in magnetostatics

Determination of the magnetic field due to current-carrying coils close to nonlinear magnetic material (saturable iron) is a particularly important application which often appears in a truly three-dimensional form not reducible to any simpler two-dimensional calculation. The problem has been formulated in terms of vector potential **A** in Chapter 2, Section 6 and the two-dimensional simplification of this formulation was discussed fully in Chapter 5. However, in three-dimensional cases the necessity for working with the three components of **A** and the difficulties now associated with the gauge specification of $\nabla \cdot \mathbf{A}$ make it desirable to look for another formulation. An approach using scalar magnetic potential has been found practical.

3.1 *Reduced scalar potential*

The basic equations,

$$\nabla \times \mathbf{H} = \mathbf{J}, \tag{3.01}$$

$$\mathbf{B} = \mu \mathbf{H}, \tag{3.02}$$

$$\nabla \cdot \mathbf{B} = 0, \tag{3.03}$$

describe the field due to a current distribution **J** near to a nonlinear

Table 6.1. *Comparison of exact eigenvalues with an appropriate matrix solution having eight degrees of freedom*

Mode number $l\,m\,n$	0 0 0	0 0 1 0 1 0 1 0 0	0 1 1 1 0 1 1 1 0	1 1 1
Exact solution $ka = (l^2 + m^2 + n^3)^{1/2}\pi$	0	3.14	4.44	5.44
ka from matrix eigenvalues	0	3.12	6.42	7.17

magnetic material of field-dependent permeability μ. The magnetic field intensity in a region containing current can be separated into two parts corresponding to a source field $\mathbf{H}_S$ and an induced magnetisation $\mathbf{H}_M$,

$$\mathbf{H} = \mathbf{H}_S + \mathbf{H}_M, \tag{3.04}$$

as in Chapter 4, Section 4. The source field is obtainable directly from the Biot–Savart law by integration over the region Ω_J containing currents:

$$\mathbf{H}_S(\mathbf{r}) = \frac{1}{4\pi} \int_{\Omega_J} \frac{\mathbf{J}(\mathbf{r}') \times (\mathbf{r} - \mathbf{r}')}{|\mathbf{r} - \mathbf{r}'|^3} \, d\Omega'. \tag{3.05}$$

This field satisfies

$$\nabla \times \mathbf{H}_S = \mathbf{J}, \tag{3.06}$$

so that

$$\nabla \times \mathbf{H}_M = 0, \tag{3.07}$$

allowing $\mathbf{H}_M$ to be written as the gradient of the *reduced scalar potential*

$$\mathbf{H}_M = -\nabla\phi. \tag{3.08}$$

In principle, this allows ϕ to be determined from the equation

$$-\nabla \cdot (\mu \nabla \phi) + \nabla \cdot (\mu \mathbf{H}_S) = 0, \tag{3.09}$$

which is obtained through Eqs. (3.02) and (3.04) by invoking the solenoidal property of $\mathbf{B}$, Eq. (3.03). In uniform, linear media Eq. (3.09) becomes

$$\nabla^2 \phi = 0, \tag{3.10}$$

since $\nabla \cdot \mathbf{H}_S = 0$ always. The effect of a discontinuity in μ at a material interface is to require the application of the field boundary rules (Chapter 2, Section 1). However, in making practical use of the reduced scalar potential ϕ for numerical problems, a difficulty, pointed out by Simkin & Trowbridge (1979), arises when materials with high permeability are involved. Within such materials the induced magnetic field $\mathbf{H}_M = -\nabla\phi$ and the source field $\mathbf{H}_S$ of Eq. (3.04) are of nearly-equal magnitude and opposite in sign, so that proportionate errors in $\mathbf{H}_M$ are magnified in the final result $\mathbf{H}$. To illustrate this the trivial problem of a 'plug' of iron having uniform permeability μ, contained within a long current-carrying solenoid (Fig. 6.8) is examined. The source field in this case for such a solenoid, carrying current I and having N turns per unit length is axial and described by the well-known result

$$\mathbf{H}_S = NI\hat{\mathbf{z}}, \tag{3.11}$$

where $\hat{\mathbf{z}}$ is a unit vector in the axial direction. If the plug fills the solenoid cross-section the field within it must be uniformly axial, so that Eq.

(3.10) has a solution

$$\phi = Cz, \tag{3.12}$$

giving, inside the magnetic plug,

$$\mathbf{H}_M = -C\hat{z}, \tag{3.13}$$

whilst elsewhere within the solenoid $\mathbf{H}_M = 0$ of course. The requirement of continuous normal $\mathbf{B}$ at the discontinuity in μ at the end faces of the plug creates a boundary condition which enables the constant C to be determined, giving

$$\mu_0 \mathbf{H}_S = \mu (\mathbf{H}_S - C\hat{z}), \tag{3.14}$$

whence

$$C = NI(1 - \mu/\mu_0). \tag{3.15}$$

The relative permeability of ferromagnetic materials is large, perhaps several thousands, in which case the cancellation of $\mathbf{H}_S$ and $\mathbf{H}_M$ in Eq. (3.04) is seen to be very close:

$$\mathbf{H} = NI\hat{z} - NI(1 - \mu/\mu_0)\hat{z}. \tag{3.16}$$

3.2 Total scalar potential

The numerical difficulties associated with the near-cancellation in Eq. (3.04) may be overcome if it is observed that in practical arrangements electric and magnetic circuits often occupy different regions of space Ω_J and Ω_K say. All of the electric currents and no magnetic materials are supposed contained within Ω_J, whilst Ω_K contains all of the magnetic materials. Within Ω_K, there being no electric currents, the field may be represented by a *total scalar potential* ψ

$$\mathbf{H} = -\nabla\psi. \tag{3.17}$$

Fig. 6.8. Plug of magnetic material in a long solenoid.

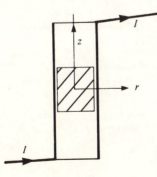

Within the current-carrying region Ω_J the reduced potential representation

$$\mathbf{H} = \mathbf{H}_S - \nabla\phi \tag{3.18}$$

must be retained, although $\mu = \mu_0$ may usually be assumed, so that ϕ is then determined by Laplace's equation, Eq. (3.10). In Ω_K, application of Eqs. (3.02) and (3.03) leads to

$$\nabla \cdot (\mu \nabla\psi) = 0. \tag{3.19}$$

Equations (3.10) and (3.19) must be solved subject to the interface requirements at the boundary between Ω_J and Ω_K, namely that normal components of $\mathbf{B}$ and tangential $\mathbf{H}$ should be continuous. Specifically, if $\mathbf{n}$ is a unit vector normal to the surface S_{JK} separating the two regions, and $\mathbf{t}$ is a unit vector tangential to the surface then

$$\mu_0(\mathbf{H}_S - \nabla\phi) \cdot \mathbf{n} = -\mu(\nabla\psi) \cdot \mathbf{n} \tag{3.20}$$

and

$$(\mathbf{H}_S - \nabla\phi) \cdot \mathbf{t} = -(\nabla\psi) \cdot \mathbf{t}. \tag{3.21}$$

It is seen that Eq. (3.21) can be integrated with respect to arc-length s to relate the potential at any two points P and Q in the interface surface S_{JK}, giving

$$\phi - \int_P^Q \mathbf{H}_S \cdot \mathbf{t} \, \mathrm{d}s = \psi, \tag{3.22}$$

where P is a point at which $\phi = \psi$ is assumed.

3.3.   *Solution of the two-potential equations*

The general nature of the problem to be considered in this section is described by Fig. 6.9. The analysis developed here follows that given by Simkin & Trowbridge (1980). It is required to determine the magnetic fields in a three-dimensional system consisting of nonlinear magnetic material (iron, perhaps), not carrying any electric current, and

Fig. 6.9. Potential regions: Q_K: iron, $\nabla \cdot (\mu\nabla\psi) = 0$; Q_J: current-carrying conductor or air, $\nabla^2\phi = 0$.

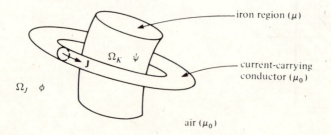

iron region (μ)

current-carrying
conductor (μ_0)

air (μ_0)

a nonmagnetic region consisting of a linear dielectric (air usually) and current-carrying conductors (copper, say). In the air-space and within the current-carrying conductors (region Ω_J) a reduced potential ϕ is used to describe the magnetic field according to Eqs. (3.04) and (3.10). It is assumed that the magnetic source field $\mathbf{H}_S$ is known, via Eq. (3.05), from a specified current distribution $\mathbf{J}$. In the saturable magnetic material (region Ω_K), because of the difficulties already described, the reduced potential representation is abandoned and total scalar potential ϕ is used. Then the field is described by Eqs. (3.18) and (3.19). The potentials ϕ and ψ are linked by means of the interface relations Eqs. (3.20) and (3.22). The complete set of relations is summarised in Table 6.2.

The problem-space is subdivided into tetrahedral elements as described and analysed in Section 2. Most of the elements will lie entirely within one or other of the regions Ω_J, Ω_K. Just a few will lie with either one vertex, one edge or one face in the dividing surface S_{JK} separating Q_J and Ω_K. It will be assumed that the tetrahedral finite elements are chosen such that each element abutting the interface S_{JK} falls into one of these three categories and that S_{JK} is accordingly modelled as the faceted surface which such an assumption implies. The problem is regarded as solved when the potential of each nodal point, ϕ or ψ as the case may be, has been determined. Notice that *both* ϕ and ψ must be determined at nodes in the interface S_{JK}.

3.4. *Solution by Galerkin's method*

At this point, in most of the cases treated in this book, a stationary functional in the field variables has been found and finite element matrix equations set up via a variational approach. Here, the

Table 6.2. *Summary of potential equations for solving three-dimensional magnetostatics problems. $\mathbf{H}_S$ is the source field, ϕ the reduced scalar potential and ψ the total scalar potential*

Region Ω_J (copper and air)	Region Ω_K (iron)
$\mathbf{H} = \mathbf{H}_S - \nabla\phi$	$\mathbf{H} = -\nabla\psi$
$\nabla^2\phi = 0$	$\nabla \cdot (\mu\nabla\psi) = 0$

Interface conditions

$$\mu_0(\mathbf{H}_S - \nabla\phi) \cdot \mathbf{n} = -\mu(\nabla\psi) \cdot \mathbf{n}$$

$$\phi - \int_P^Q \mathbf{H}_S \cdot \mathbf{t} \, ds = \psi$$

complications of the interface conditions linking Ω_J and Ω_K point to the use instead of the *method of weighted residuals*, a powerful alternative which is sometimes employed in setting up finite element matrix equations (see Zienkiewicz, 1976). The method invokes the partial differential equations and the boundary constraints of the problem directly and hence is applicable to cases where a stationary functional is not obvious. In instances where a stationary functional does exist, the *Galerkin* option of the general method of weighted residuals (see Chapter 4, Section 2.1) gives exactly the same result as the variational approach.

As previously in the variational method, the field variable u (say) is approximated by means of interpolation functions α_m. Within a finite element, or assembly of such, an approximate solution U is written as

$$U = \sum_{m=1}^{N} U_m \alpha_m, \tag{3.23}$$

in terms of the nodal potentials U_m. The task, as before, is to find U_m such that, within the limitations of the chosen functions α_m, Eq. (3.23) represents a 'best fit' to the problem differential equations and boundary conditions.

The method of weighted residuals can be applied to problems described by the general equation Eq. (2.03), but for the two-potential problem as formulated in this section, it is sufficient to consider

$$\nabla \cdot (p\nabla u) = 0, \tag{3.24}$$

representing either Eq. (3.10) or Eq. (3.19). Leaving aside for the moment the interface conditions, Eqs. (3.20) and (3.22), the boundary conditions might be comprised of the Neumann condition

$$p\frac{\partial u}{\partial n} = q \tag{3.25}$$

on surface S_1, with q some given function of position, and the Dirichlet condition

$$u = u_0 \tag{3.26}$$

on surface S_2, with u_0 also a function of position. The problem-space is supposed to be the volume Ω enclosed by the closed surface comprising of S_1 and S_2.

A set of arbitrary *weighting functions* W_n, $\overline{W}_n$ and $\overline{\overline{W}}_n$ is selected and *weighted residuals*

$$R_n = \int_\Omega W_n \nabla \cdot (p\nabla U)\, \mathrm{d}\Omega + \int_{S_1} \overline{W}_n \left(p\frac{\partial U}{\partial n} - q\right) \mathrm{d}S$$
$$+ \int \overline{\overline{W}}_n (U - u_0)\, \mathrm{d}S \tag{3.27}$$

are defined for the approximation (3.23). Clearly, if U were the exact solution of Eq. (3.24) the residuals R_n would vanish under all circumstances. But U in Eq. (3.23) is an approximation, with N undetermined coefficients U_m. If N linearly independent weightings are chosen for Eq. (3.27) there will be enough equations to determine the set U_m by putting each weighted residual R_n to zero. Then U of Eq. (3.23) may be regarded as a 'best fit' for the particular class of weighting functions chosen.

Equation (3.27) may be cast into a form which does not specifically require second derivatives of U to be defined. This is done by applying the vector calculus identity

$$\nabla \cdot (p W_n \nabla U) = W_n \nabla \cdot (p \nabla U) + p (\nabla W_n) \cdot (\nabla U) \qquad (3.28)$$

and the divergence theorem

$$\int_\Omega \nabla \cdot (p W_n \nabla U) = \int_S W_n p \frac{\partial U}{\partial n} \, dS, \qquad (3.29)$$

where S is the surface which encloses the volume Ω. Then, since S is comprised of S_1 and S_2, Eq. (3.27) becomes

$$R_n = -\int_\Omega p(\nabla W_n) \cdot (\nabla U) \, d\Omega + \int_{S_1} (W_n + \overline{W}_n) p \frac{\partial U}{\partial n} \, dS$$

$$- \int_{S_1} \overline{W}_n q \, dS + \int_{S_2} W_n p \frac{\partial U}{\partial n} \, dS + \int_{S_2} \overline{W}_n (U - u_0) \, dS. \qquad (3.30)$$

In most cases q may be taken as zero (homogeneous Neumann), whilst U is constrained such that $U = u_0$ on S_2. No loss of generality is incurred by choosing $W_n = -\overline{W}_n$, so that a relatively simple relationship

$$R_n = -\int_\Omega p(\nabla W_n) \cdot (\nabla U) \, d\Omega + \int_{S_2} W_n p \frac{\partial U}{\partial n} \, dS \qquad (3.31)$$

ensues. The residual R_n in Eq. (3.31) is now put to zero. This operation sets up matrix equations, either (say)

$$\mathbf{S}_\phi \mathbf{\Phi} = 0 \qquad (3.22)$$

or

$$\mathbf{S}_\psi \mathbf{\Psi} = 0, \qquad (3.33)$$

for each element not containing nodes in the interface S_{JK} between Ω_J and Ω_K, where the symbols $\mathbf{\Phi}$ and $\mathbf{\Psi}$ represent vectors of nodal potentials to be determined. Notice that for an element completely 'buried' in the space either Ω_J or Ω_K, the surface S_2 might be considered to be an interface between two adjacent elements. The material property p (here in fact permeability μ) is allowed to be discontinuous between elements, in which case nevertheless $p \partial u / \partial n$, the *normal flux density*, must be continuous. The common surface S_2 between two elements will necessarily have equal and opposite normal vectors **n**. Hence, the second term in Eq. (3.32) should cancel when the residuals for adjacent elements are combined. Thus, *neglecting* the second term of Eq. (3.31), contribution over element interfaces is equivalent to *enforcing* continuity of $p\partial u/\partial n$. This being so it may be further noted that the surface-integral term in Eq. (3.31) could therefore only enter into the analysis when referring to an external Dirichlet boundary. But U is constrained to be equal to u_0 on external boundaries S_2, so there is no need to evaluate residuals at such boundary points. Thus, in practical calculations, the second term of Eq. (3.31) is ignored completely. Clearly, the matrices $\mathbf{S}_\phi$ and $\mathbf{S}_\psi$ connect in the way which has already been described in Chapter 1, Section 5, for an assembly of finite elements.

If now the arbitrary weighting functions W_n are in fact chosen to be the same as the interpolation functions α_m used in Eq. (3.23), the analysis here becomes what is known as the *Galerkin* procedure. The method is then equivalent to the variational approach and, consequently, for elements not in the interface, the **S**-matrices are determined by

$$S_{mn} = \sum_{i=1}^{3} \sum_{j=i+1}^{4} pK_{ij}Q_{mn}^{ij}, \tag{3.34}$$

exactly as in Eq. (2.31). Notice that **T**-matrices, such as occur in Eq. (2.19) are absent here, corresponding to the absence of a right-hand side term in either of Eqs. (3.10) and (3.19). There is no particular difficulty in introducing such terms, as would be required, for instance, when considering nonlinear material in Ω_J or permanent-magnet material in Ω_K (see Simkin & Trowbridge, 1980). It may be observed that the absence here of terms due to any homogeneous Neumann boundary constraint corresponds to what was described earlier in the variational analyses as a 'natural' boundary condition.

Now consider two finite elements Ω_1 and Ω_2 lying, respectively, in Ω_J and Ω_K and having common faces in the interface S_{12} between Ω_J and Ω_K. Substituting $p=1$ and $p=\mu/\mu_0\mu_r$ in the ϕ- and ψ-expressions respectively, residuals

$$R_{Jn} = -\int_{\Omega_1} (\nabla W_n) \cdot (\nabla\phi)\, d\Omega + \int_{S_{12}} W_n \frac{\partial\psi}{\partial n}\, dS \tag{3.35}$$

and

$$R_{Kn} = -\int_{\Omega_2} \mu_r(\nabla W_n) \cdot (\nabla\psi)\, d\Omega + \int_{S_{21}} W_n \frac{\partial\psi}{\partial n}\, dS \tag{3.36}$$

may be defined, where S_{12} and S_{21} are essentially the same surface but have equal and opposite normal vectors **n**. The surface S_{12} (and S_{21}) is considered to be part of a surface of the type S_2. Since it is to cancel anyway, the surface integral over the remainder of surface S_2, buried in Ω_J (or Ω_K) as it is, need not be included in the expression for R_{Jn} (or R_{Kn}). The sum of the residuals

$$R_{Jn} + R_{Kn} = -\int_{\Omega_1} (\nabla W_n) \cdot (\nabla \phi) \, d\Omega - \int_{\Omega_2} \mu_r (\nabla W_n) \cdot (\nabla \psi) \, d\Omega$$

$$+ \int_{S_{12}} W_n \left(\frac{\partial \phi}{\partial n} - \mu_r \frac{\partial \psi}{\partial n} \right) dS \tag{3.37}$$

may be rewritten using the interface relationship Eq. (3.20) as

$$R_{Jn} + R_{Kn} = -\int_{\Omega} (\nabla W_n) \cdot (\nabla \phi) \, d\Omega - \int_{\Omega_2} \mu_r (\nabla W_n) \cdot (\nabla \psi) \, d\Omega$$

$$+ \int_{S_{12}} W_n \mathbf{H_S} \cdot n \, dS. \tag{3.38}$$

It is clear that setting the residual in Eq. (3.38) above to zero yields a matrix equation of form

$$\mathbf{S}'_\phi \mathbf{\Phi} + \mathbf{S}'_\psi \mathbf{\Psi} = \mathbf{h}, \tag{3.39}$$

where the **S**-matrices of Eq. (3.39) have the same form as Eqs. (3.32) and (3.33) when relating to a completely buried element. The vector **h** is formed by surface integration over the common interface of the elements Ω_1 and Ω_2,

$$h_n = \int_{S_{12}} W_n \mathbf{H_S} \cdot \mathbf{n} \, dS. \tag{3.40}$$

(Here the index n and the unit vector **n** must not be confused.) There is always ambiguity as to which of ϕ and ψ represents the field at the interface. This is resolved by eliminating one or other using the relationship (3.32), having arbitrarily chosen one point at which $\phi = \psi$ is assumed. Once again, the Galerkin option may be chosen by putting $W_n = \alpha_n$. Now the finite element matrices $\mathbf{S}'_\phi$ and $\mathbf{S}'_\psi$ may be 'connected' to the matrices corresponding to elements not having a surface in the interface, as ordinarily with stationary functional-derived arrays. Notice that if an element merely has a single vertex, or an edge in the surface S_{12}, the integral of Eq. (3.40) vanishes whereas, nevertheless, Eq. (3.22) is still significant. Having set up the global matrices $\mathbf{S}_\phi$ and $\mathbf{S}_\psi$, these of course represent a set of linear equations in the unknown nodal potentials ϕ and ψ. The unknowns may be solved for and subsequently processed to determine fields, forces and so forth. Thus, practical three-dimensional problems involving current specified in conductors producing magnetic fields in neighbouring saturated materials may be tackled.

The procedure outlined here is the essence of the TOSCA program described by Simkin & Trowbridge (1980). It has been shown to produce useful results for some three-dimensional nonlinear magnetic problems in magnetostatics which hitherto had not been solvable by computer methods. The reader should consult the original paper by Simkin & Trowbridge for more details of this rather complex analysis and for a presentation of the practical results which were obtained.

4.   Three-dimensional problems in electromagnetic-wave propagation

In this section the problem considered is that of finding the steady-state time-harmonic electromagnetic fields at frequency ω in a waveguide or cavity containing arbitrarily-shaped obstacles of perfect conductors or lossless, linear, isotropic dielectric and magnetic materials. The system is supposed as being bounded by perfect conductors except at certain specific surfaces where the tangential fields are prescribed. The complete bounding surface is of course closed.

Problems of this nature arise in determining the scattering parameters of obstacles in waveguides and in the analysis of microwave cavities. In many cases a two-dimensional treatment is possible but sooner or later geometries arise for which a full three-dimensional analysis is the only avenue. Although the possibility of using the potentials ϕ and $\mathbf{A}$ is not discounted, the complications of applying three-dimensional boundary constraints to these variables point to the use of the field vectors $\mathbf{H}$ and $\mathbf{E}$ directly. In such cases the functionals numbered (4.09) and (4.19) in Chapter 2 may be shown to lead directly to

$$F(\mathbf{H}) = \tfrac{1}{2} \int_{\Omega} [(\nabla \times \mathbf{H})^2 / \varepsilon_r - k^2 \mu_r \mathbf{H}^2] \, d\Omega, \tag{4.01}$$

$$F(\mathbf{E}) = \tfrac{1}{2} \int_{\Omega} [k^2 \varepsilon_r \mathbf{E}^2 - (\nabla \times \mathbf{E})^2 / \mu_r] \, d\Omega, \tag{4.02}$$

where normalised angular frequency $k = \omega (\mu_0 \varepsilon_0)^{1/2}$ and relative constitutive constants $\varepsilon_r = \varepsilon / \varepsilon_0$, $\mu_r = \mu / \mu_0$ are used. (The fine distinction between $\mathbf{H}$ and $\mathbf{E}$, satisfying Maxwell's equations, and $\mathbf{H}'$, $\mathbf{E}'$, trial functions adjusted to make the functional stationary, has been dropped.)

Equations (4.01) and (4.02) were derived in Chapter 2 as valid functionals implicitly making the very wide assumption of complex μ_r and ε_r, corresponding to the occurrence of internal dissipation and power loss. In fact even complex k could be considered, corresponding to the implied $\exp(j\omega t)$ phasor factor decaying or growing with time. However, many problems are concerned with the lossless case where μ_r, ε_r and k

are entirely real. Then it is clear that the real and imaginary parts of $\mathbf{H} = \mathbf{H}_r + j\mathbf{H}_i$ each satisfy the equation

$$\nabla \times \nabla \times \mathbf{H} = k^2 \mu_r \varepsilon_r \mathbf{H} \qquad (4.03)$$

separately. The analysis showing that $F(\mathbf{H})$ is stationary (Chapter 2, Section 4) hinges upon Eq. (4.03) being satisfied subject to boundary constraints. Thus $F(\mathbf{H}_r)$ and $F(\mathbf{H}_i)$ separately are each stationary about the correct electromagnetics solution. Since

$$\mathbf{H} \cdot \mathbf{H}^* = \mathbf{H}_r^2 + \mathbf{H}_i^2, \qquad (4.04)$$

where $\mathbf{H}^*$ is the complex conjugate of $\mathbf{H}$, it is evident that the functional

$$F(\mathbf{H} \cdot \mathbf{H}^*) = \tfrac{1}{2} \int_{\Omega} [(\nabla \times \mathbf{H}) \cdot (\nabla \times \mathbf{H}^*)/\varepsilon_r - k^2 \mu_r \mathbf{H} \cdot \mathbf{H}^*] \, d\Omega \quad (4.05)$$

is also stationary provided the lossless case is being considered. It may also be shown that

$$F(\mathbf{H}_r \cdot \mathbf{H}_i) = \tfrac{1}{2} \int_{\Omega} [(\nabla \times \mathbf{H}_r) \cdot (\nabla \times \mathbf{H}_i)/\varepsilon_r - k^2 \mu_r \mathbf{H}_r \cdot \mathbf{H}_i] \, d\Omega \quad (4.06)$$

is a valid functional in the lossless case.

Similar arguments give electric field functionals alternative to Eq. (4.02). The functionals (4.05) and its electric counterpart have been used by Ferrari & Maile (1978) to set up a finite element system solving waveguide discontinuity problems using first-order tetrahedral elements. A broadening and extension of this work, using the more general functionals (4.01) and (4.02) and for higher-order tetrahedra, will be developed here. The variable $\mathbf{H}$ is chosen as the working unknown, but when it has been determined by the approximate methods of finite element analysis the vector $\mathbf{E}$, if required, can clearly be evaluated in similar approximation from the Maxwell equation

$$\nabla \times \mathbf{H} = j\omega\varepsilon \, \mathbf{E}. \qquad (4.07)$$

An analogous development treating $\mathbf{E}$ as the working unknown may be given. It turns out that the final computer programs can be easily adapted for either working variable.

4.1   *The $\mathbf{H}$-solution*

A closed surface S, enclosing the volume Ω of the integration (4.01) and made up from the surfaces S_1 and S_2 is considered. On S_1 there is specified a generalised homogeneous Neumann condition

$$(\nabla \times \mathbf{H}) \times \mathbf{n} = 0, \qquad (4.08)$$

which, by virtue of Eq. (4.07) and the boundary rules for electric fields, corresponds to a perfectly conducting wall or short circuit. On S_2 it is supposed that either a generalised Dirichlet condition holds, either homogeneous,

$$\mathbf{H} \times \mathbf{n} = 0, \tag{4.09}$$

or inhomogeneous,

$$\mathbf{H} \times \mathbf{n} = \mathbf{H}_p. \tag{4.10}$$

The homogeneous condition of tangential $\mathbf{H}$ vanishing, corresponds to planes of symmetry in an electric field (open circuit) whilst its inhomogeneous counterpart relates to the specification of tangential $\mathbf{H}$ on surfaces or 'windows' of ports coupled to the outside world. It was seen in Chapter 2 that if a closed surface S is made up piecewise from surfaces such as S_1 and S_2, then the functional $F(\mathbf{H})$ of Eq. (4.01) is stationary. In practical terms this means that in any finite element analysis the three components of $\mathbf{H}$ at any perfect conductor boundary may be left free to vary without any constraint whatsoever, whilst at Dirichlet boundaries the two tangential components must be specified, only the single normal component being free to vary. The volume Ω is broken into tetrahedra in each of which the material properties ε_r and μ_r are constant, although they may be discontinuous across element boundaries.

4.2   *The expressions for a single tetrahedron*

In each tetrahedron the field is approximated to order n_0 by

$$\mathbf{H} = \sum_{m=1}^{N} \mathbf{H}^m \alpha_m(\boldsymbol{\zeta}), \tag{4.11}$$

where $\boldsymbol{\zeta} = (\zeta_1, \zeta_2, \zeta_3, \zeta_4)$ is the three-dimensional homogeneous coordinate vector of Section 2.1, whilst the $\alpha_m(\boldsymbol{\zeta})$ are single-index versions of the interpolation polynomials described in Sections 2.2 and 2.3. As before, m indexes the $N = (n_0+1)(n_0+2)(n_0+3)/6$ nodal points of the volume V meshed for a polynomial approximation of order n_0. Purely in order to clear space in the notational jungle, some subscripts have been elevated. Such an elevation does not signify an exponent. Another piece of dead wood which might hamper view is the summation sign such as in Eq. (4.11). Henceforth, the occurrence of a repeated index such as m in Eq. (4.11) will imply summation over the whole of its permissible range (in that case $m = 1, 2, \ldots, N$). It will be recalled from Sections 2.2 and 2.3 that the polynomials α_m are such that the trial function for $\mathbf{H}$, Eq. (4.11), automatically yields the unknown vector parameter $\mathbf{H}^m$ at each node m. Equation (4.11) may be substituted into Eq. (4.01) to give the functional corresponding to just one element,

$$F = \tfrac{1}{2} H_s^m H_t^n \left(\frac{V}{\varepsilon_r} \hat{Q}_{mn}^{ij} K_{ij}^{st} - V \mu_r k^2 \delta_{st} T_{mn} \right), \tag{4.12}$$

where H_s^m is the sth Cartesian component at node m,

$$\hat{Q}_{mn}^{ij} = 6 \int_\Omega \frac{\partial \alpha_m}{\partial \zeta_i} \frac{\partial \alpha_n}{\partial \zeta_j} \, d\zeta_1 \, d\zeta_2 \, d\zeta_3, \tag{4.13}$$

$$T_{mn} = 6 \int_\Omega \alpha_m \alpha_n \, d\zeta_1 \, d\zeta_2 \, d\zeta_3, \tag{4.14}$$

$$K_{ij}^{st} = \delta_{st} \frac{\partial \zeta_i}{\partial r_w} \frac{\partial \zeta_j}{\partial r_w} \frac{\partial \zeta_i}{\partial r_t} \frac{\partial \zeta_j}{\partial r_s}. \tag{4.15}$$

$m, n = 1, 2, \ldots, N$ (tetrahedron node numbering,
$$N = (n_0 + 1)(n_0 + 2)(n_0 + 3)/6),$$

$s, t, w = 1, 2, 3$ (Cartesian axis labelling),

$i, j = 1, 2, 3, 4$ (tetrahedron vertex numbering),

$\delta_{st} = 1$ if $s = t$, $\delta_{st} = 0$ if $s \neq t$ (Kronecker delta).

The integrations (4.13) and (4.14) have been designated as being with respect to $\zeta_1, \zeta_2, \zeta_3$, it being implied that $(1 - \zeta_1 - \zeta_2 - \zeta_3)$ replaces ζ_4 wherever it occurs. Of course, which of the ζs is eliminated in this fashion is in fact entirely arbitrary. The result here follows the pattern set in the single variable case and is written in terms of:

(i) two matrices $\hat{Q}$ and T which are independent of the tetrahedron geometry apart from scale factors and may be computed once and for all.

(ii) an array K, this time a 12×12 matrix, which must be computed for each element but which is a simple function of the Cartesian coordinates of the tetrahedron vertices, involving no volume integration.

The matrix T has been encountered before in Section 2. The matrix $\hat{Q}$ is obviously closely related to the Q-matrix of Section 2 but some of the symmetries of the simpler case there do not now apply. Consequently, the slightly more general form $\hat{Q}$ is retained. Matrices $\hat{Q}$ and T up to order 4 have been computed and successfully used in waveguide analyses.

Using Eq. (4.12), a relation may be written in terms of a symmetric matrix W and a column vector H_c of the tetrahedron nodal field components:

$$F = \tfrac{1}{2} H_c^T W H_c. \tag{4.16}$$

4.3. Construction of the global matrices

As before, the simple quadratic form of Eq. (4.16) allows the functional for a connected set of tetrahedral elements to be written as

$$F_{con} = \tfrac{1}{2} H_{con}^T C^T W_{dis} C H_{con}. \tag{4.17}$$

Thus, Eq. (4.16) serves to represent both the single tetrahedron and the connected set, provided in the latter case that $\mathbf{W}$ represents a global matrix properly constructed from individual tetrahedron arrays, whilst $\mathbf{H}_c$ is the vector representing all the assembled tetrahedron nodal field components. As might be expected intuitively, by analogy to two dimensions, the relative permeability μ_r and the relative permittivity ε_r, although assumed constant within each tetrahedron, may be discontinuous on going from one tetrahedron element to another. This permits space-varying μ_r and ε_r to be accurately modelled provided the tetrahedron volumes are sufficiently small.

The boundary constraints described earlier must now be imposed upon the global matrix relationship (4.16). As was explained in Section 4.1, the natural, homogeneous Neumann boundary condition corresponding to perfect conductors needs no specific constraints upon the boundary variables. However, the Dirichlet condition, corresponding either to a 'window' coupling to the outside world (inhomogeneous Dirichlet) or to a plane of symmetry (homogeneous Dirichlet) requires the constraint of components of $\mathbf{H}$ in the boundary plane. To simplify this constraint, the Dirichlet surface is allowed to consist only of a number of planes, each perpendicular to one of the three Cartesian axes of the problem. Then prescribing $\mathbf{H} \times \mathbf{n} = \mathbf{P}$ on, for example, a plane perpendicular to the z-axis, involves setting $H_x = -P_y$, $H_y = P_x$ and leaving H_z free to vary. The matrix equation, Eq. (4.16) then partitions simply in the usual way:

$$F = \tfrac{1}{2}\mathbf{H}_f^T\mathbf{W}_{ff}\,\mathbf{H}_f + \mathbf{H}_f^T\mathbf{W}_{fp}\,\mathbf{H}_p + \tfrac{1}{2}\mathbf{H}_p^T\mathbf{W}_{pp}\mathbf{H}_p, \qquad (4.18)$$

where $\mathbf{H}_f$ and $\mathbf{H}_p$ are, respectively, the column vectors of the free and prescribed components of $\mathbf{H}$. Finally, to find an appropriate solution to the problem, the functional F is made stationary with respect to all variations of the vector $\mathbf{H}_f$. This leads to the linear matrix equation

$$\mathbf{W}_{ff}\mathbf{H}_f = -\mathbf{W}_{fp}\,\mathbf{H}_p, \qquad (4.19)$$

which may be solved for the unknown vector $\mathbf{H}_f$.

4.4. Application to waveguide problems

An application of the method just described is to be found in determining the scattering due to obstacles in waveguide arrangements. A typical geometry is illustrated schematically in Fig. 6.10. Two hollow cylindrical waveguides, with perfectly conducting walls but arbitrary and different cross-sections, meet in an irregularly-shaped region containing linear, isotropic, sourceless materials. The problem is to compute the scattering matrix of the junction region with respect to dominant mode waveguide fields at reference planes R_1 and R_2.

The scattering matrix S of a multi-port microwave device is defined in terms of normalised complex coefficients a and b representing incident and reflected wave amplitudes in each of the waveguides feeding the device. In the case of a lossless two-conductor transmission line of characteristic impedance Z_{0n}, connected to the nth port of a multi-port arrangement, the normalised amplitudes are defined precisely in terms of the voltages or currents of incident and reflected waves as

$$a_n = \frac{V_{n+}}{\sqrt{Z_{0n}}} = \sqrt{(Z_{0n})}I_{n+}, \tag{4.20}$$

$$b_n = \frac{V_{n-}}{\sqrt{Z_{0n}}} = -\sqrt{(Z_{0n})}I_{n-}. \tag{4.21}$$

Evidently, from the ordinary rules of circuit analysis, the average power flowing into the nth port is

$$W_n = \tfrac{1}{2}(a_n a_n^* - b_n b_n^*). \tag{4.22}$$

In such a case the definition of (say) a two-port S presents no problems:

$$\begin{bmatrix} b_1 \\ b_2 \end{bmatrix} = \begin{bmatrix} S_{11} & S_{12} \\ S_{21} & S_{22} \end{bmatrix} \begin{bmatrix} a_1 \\ a_2 \end{bmatrix}. \tag{4.23}$$

The scattering matrix S is seen to be a generalised voltage reflection coefficient. The theory of scattering matrices with respect to two-conductor lines is developed fully in standard works on microwaves, such as Ramo, Whinnery & Van Duzer (1965) or Collin (1966). Waveguides are but another example of high-frequency transmission lines. Thus, although developed from a different standpoint, scattering parameter analysis and applications may freely be applied to waveguides.

Fig. 6.10. A typical 2-port waveguide scattering problem.

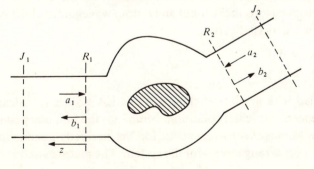

Since standing wave measurements of reflection coefficients are easily made in waveguides, there is no particular difficulty in determining **S** experimentally. However, if **S** is to be calculated from the geometry and physical properties of the n-port it soon becomes apparent that in waveguides there are no uniquely defined transmission-line voltages or currents. Instead, the normalised wave amplitudes a and b are defined from the transverse waveguide fields. To obtain unique results, the coefficients have to relate to a specific mode, usually the dominant one. Real vectors $\mathbf{e}(x, y)$ and $\mathbf{h}(x, y)$, proportional to the transverse electric and magnetic fields propagating in the z-direction, are defined such that

$$\int_{P} (\mathbf{e} \times \mathbf{h}) \cdot dS = 1, \tag{4.24}$$

$$\frac{|\mathbf{e}|}{|\mathbf{h}|} = Z_{\mathrm{W}}, \tag{4.25}$$

where P is the plane perpendicular to the waveguide axis (z-axis) and Z_{W} is the *wave impedance* of the mode (see for instance Ramo, Whinnery & Van Duzer, 1965, pp. 404 and 409). The actual waveguide transverse fields in any given case may be represented as

$$\mathbf{E}_{t} = (b \exp(-\mathrm{j}\beta z) + a \exp(\mathrm{j}\beta z)]\mathbf{e}, \tag{4.26}$$

$$\mathbf{H}_{t} = [b \exp(-\mathrm{j}\beta z) - a \exp(\mathrm{j}\beta z)]\mathbf{h}, \tag{4.27}$$

with respect to reference planes designated R_1 and R_2 in Fig. 6.10. It is observed that the average power flow in the waveguide obtained by integrating the Poynting vector over the waveguide cross-section

$$W = \tfrac{1}{2} \int_{P} (\mathbf{E}_{t} \times \mathbf{H}_{t}^{*}) \cdot d\mathbf{S} \tag{4.28}$$

which becomes

$$W = \tfrac{1}{2}(aa^{*} - bb^{*}), \tag{4.29}$$

agreeing with Eq. (4.22)

A general procedure for determining **S** now emerges. At two planes, J_1 and J_2 respectively, in the input and output waveguides of the two-port depicted in Fig. 6.10, fields

$$\mathbf{H}_{t1} = A_1 \mathbf{h}, \tag{4.30}$$

$$\mathbf{H}_{t2} = A_2 \mathbf{h}, \tag{4.31}$$

are specified with arbitrarily chosen A_1 and A_2. Such specification, an inhomogeneous Dirichlet condition, completes the boundary constraints (otherwise homogeneous Neumann for the perfectly conducting walls of the two-port arrangement) for the system. The matrices and prescribed vectors of Eq. (4.19) can be set up and the free-to-vary field components H_f calculated. Comparison of Eq. (4.27) and the computed fields near planes J_1 and J_2 allows the reflection coefficients b_1/a_1 and b_2/a_2 to be determined. It is assumed that the planes J_1 and J_2 are each sufficiently far from the inhomogeneous region for all evanescent fields to have decayed to negligible amplitudes. A little algebraic manipulation of Eq. (4.23) reveals that

$$S_{21}S_{12} = \left(\frac{b_1}{a_1} - S_{11}\right)\left(\frac{b_2}{a_2} - S_{22}\right). \tag{4.32}$$

If the materials of the two-port are all isotropic, then $S_{12} = S_{21}$. Evidently, three independent determinations of b_1/a_1 and b_2/a_2 are sufficient to evaluate S_{11}, S_{22} and S_{12}.

Although cases where the inhomogeneous region is lossy can be dealt with by the method developed here, the present analysis will be restricted to the lossless case. It is observed that with such restriction the matrices of Eq. (4.19) are entirely real. However, even in the lossless case, the fields in Eq. (4.19) are in general complex. Thus the arithmetic treatment of that equation would in general be complex in order to solve a typical lossless field problem. However, if somewhere in one of the waveguides of Fig. 6.10 there exists a short circuit or open circuit, then the power flowing in the waveguide is reflected completely. The net power flow is zero so that $aa^* = bb^*$ everywhere. There exists in the connected waveguide system a standing wave which would be recognised in either uniform section as having an *infinite standing wave ratio*. It is easily shown that in such cases the fields (4.26) and (4.27) may be represented by purely real numbers. Evidently, there is a simplicity in assuming one of the 'window' planes, say J_1, to be an open circuit, $H_t = 0$, $b_1/a_1 = 1$ or a short circuit, $E_t = 0$, $b_1/a_1 = -1$ (see Eqs. (4.26) and (4.27)) so that one of the ratios b/a required to determine S is known in advance. The three required values of the reflection coefficient b_2/a_2 may be obtained by changing the position of the short or open circuit plane J_1 to each of three different alternatives but keeping J_2 fixed. Note that this is very similar to a standard experimental technique for finding scattering parameters (see for instance Sucher & Fox, 1963). Instead of determining b_2/a_2 at J_2, another possibility is, instead, to vary the position of J_2 for each of the three chosen J_1. With a fixed prescribed H_{t2}, in each of the three cases, a position J_2' will eventually be found for which the vector solution H_f of Eq. (4.19) is infinite. It corresponds to finding a nodal plane ($H_t = 0$) in the lossless cavity formed by the short circuiting of plane J_1. Let the position of the nodal plane J_2' be $z = s$ (see Fig. 6.11).

Whether or not J_2' has been located physically it nevertheless exists (obviously it must be assumed that the waveguide feeding port 2 is sufficiently long). Then $\mathbf{H}_t = 0$ at $z = s$ so that, at J_2, corresponding to Eq. (4.27),

$$\mathbf{H}_{t2} = A_2[\exp -\mathrm{j}\beta(z-s) - \exp \mathrm{j}\beta(z-s)]\mathbf{h}, \tag{4.33}$$

$$= -2\mathrm{j}A_2 \sin \beta(z-s)\mathbf{h}. \tag{4.34}$$

(Since the field amplitudes are taken to be purely real everywhere, it is clear that the arbitrary amplitude A_2 has in fact been chosen to be imaginary.) Evidently, the required reflection coefficient at plane J_2 is, from Eq. (4.33),

$$b_2/a_2 = \exp(2\mathrm{j}\beta s). \tag{4.35}$$

Either s is found by the quasi-resonance method directly or else the computed fields at mesh points to the left of J_2 are examined and a value s is found for a fit with Eq. (4.34).

4.5.   *Calculation of S from the functional*

Two rather obvious methods for determining the reflection coefficients of the short or open circuited waveguide system have been described. A third method, not so readily apparent, uses the functional $F(\mathbf{H}')$ itself. It merits discussion on its own since it embodies a principle which is of paramount importance in finite element analyses deriving from the stationarity of a functional. $F(\mathbf{H}')$, whilst being approximate, since it is built up piecewise from simple trial functions within each element, nevertheless is likely to be a very good estimate of $F(\mathbf{H})$, the true value of F. This is because of the very stationarity of F; if $\mathbf{H}' = \mathbf{H} + \delta\mathbf{H}$ describes the precision of the finite element solution, then $F(\mathbf{H}')$ is accurate to the order $\delta\mathbf{H}^2$:

$$F(\mathbf{H}') = F(\mathbf{H}) + 0(\delta\mathbf{H}^2). \tag{4.36}$$

Put in somewhat naive terms, $F(\mathbf{H}')$ and any quantity derivable from it is averaged from *all* the nodes employed in the system, whereas methods

Fig. 6.11. Diagram showing the reference plane R_2, an output plane J_2 and a short circuit plane J_2', either located or implied.

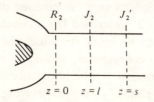

which merely use $\mathbf{H}'$ itself derive parameters from the averages of just a few nodes.

Now examine the value of the functional

$$F(\mathbf{H}) = \tfrac{1}{2} \int_{\Omega} [(\nabla \times \mathbf{H})^2 / \varepsilon_r - k^2 \mu_r \mathbf{H}^2] \, d\Omega. \tag{4.37}$$

Use the vector identity

$$(\nabla \times \mathbf{H})^2 = \nabla \cdot (\mathbf{H} \times \nabla \times \mathbf{H}) + \mathbf{H} \cdot (\nabla \times \nabla \times \mathbf{H}), \tag{4.38}$$

so that Eq. (4.37) becomes

$$F(\mathbf{H}) = \tfrac{1}{2} \int_{\Omega} \left[\mathbf{H} \cdot \left(\frac{\nabla \times \nabla \times \mathbf{H}}{\varepsilon_r} - k^2 \mu_r \mathbf{H} \right) + \frac{\nabla \cdot (\mathbf{H} \times \nabla \times \mathbf{H})}{\varepsilon_r} \right] d\Omega. \tag{4.39}$$

Remembering that $(\nabla \times \nabla \times \mathbf{H}/\varepsilon_r - k^2 \mu_r \mathbf{H})$ vanishes for $\mathbf{H}$, satisfying Maxwell's equations, and also applying the divergence theorem, then gives

$$F(\mathbf{H}) = \tfrac{1}{2} \int_{S} \frac{\mathbf{H} \times (\nabla \times \mathbf{H})}{\varepsilon_r} \cdot d\mathbf{S}. \tag{4.40}$$

Using the Maxwell equation

$$\nabla \times \mathbf{H} = j\omega \varepsilon_0 \varepsilon_r \mathbf{E} \tag{4.41}$$

then gives

$$F(\mathbf{H}) = \tfrac{1}{2} j\omega \varepsilon_0 \int_{S} (\mathbf{H} \times \mathbf{E}) \cdot \mathbf{n} \, dS = \tfrac{1}{2} j\omega \varepsilon_0 \int_{S} (\mathbf{E} \times \mathbf{n}) \cdot \mathbf{H} \, dS. \tag{4.42}$$

Now for the cavity problem (port 1 of the waveguide system, either a short circuit, $\mathbf{E} \times \mathbf{n} = 0$, or an open circuit, $\mathbf{H} = 0$), it is clear that the surface integral of Eq. (4.42) vanishes everywhere except over the input plane J_2. At the input plane of port 2, from Eqs. (4.26) and (4.27) writing $(b_2/a_2) \exp(-2j\beta l) = \rho$,

$$\mathbf{E}_t = a_2 \, e^{j\beta l} (\rho + 1)\mathbf{e}, \tag{4.43}$$

$$\mathbf{H}_t = a_2 \, e^{j\beta l} (\rho - 1)\mathbf{h}, \tag{4.44}$$

so that

$$F(\mathbf{H}) = \tfrac{1}{2} j\omega \varepsilon_0 a_2^2 \, e^{2j\beta l} (\rho + 1)(\rho - 1) \int_{J_2} (\mathbf{e} \times \mathbf{h}) \cdot \mathbf{n} \, dS. \tag{4.45}$$

But the Dirichlet condition specified over J_2 was

$$\mathbf{H}_t = A_2 \mathbf{h}. \tag{4.46}$$

By definition, $\int (\mathbf{e} \times \mathbf{h}) \cdot \mathbf{n} \, dS = 1$, so using Eq. (4.44), a_2^2 can be eliminated and Eq. (4.45) becomes

$$F(\mathbf{H}) = \tfrac{1}{2} j\omega \varepsilon_0 \left(\frac{\rho + 1}{\rho - 1} \right) A_2^2. \tag{4.47}$$

As has been already pointed out, a very good estimate of $F(\mathbf{H})$ is available in $F(\mathbf{H}')$, coming from a back substitution of $\mathbf{H}'$ into the equation

$$F(\mathbf{H}') = \tfrac{1}{2}\mathbf{H}'^{\mathrm{T}}\mathbf{W}\mathbf{H}'. \tag{4.48}$$

Thus a good estimate of ρ, hence a_2/b_2, is available from Eq. (4.46). Then, as previously explained, moving the shorting plane J_1 to three different locations enables the scattering matrix $\mathbf{S}$ to be calculated.

4.6.   *Practical example – metal sphere in a waveguide*

A large-scale computer program has been written, which implements the numerical method described here. The data input required by the program is normalised frequency $k = \omega(\mu_0\varepsilon_0)^{1/2}$ overall and, for each element separately, μ_r and ε_r. Geometrical information is supplied by breaking up the volume of the problem into hexahedra (eight-node bricks) and pentahedra (six-node). The program itself divides these into tetrahedra (five per hexahedron and three per pentahedron – see Fig. 6.12). Boundary information must be given wherever this is homogeneous or inhomogeneous Dirichlet. The matrix equation, Eq. (4.19), is solved and there is provision for evaluation of $F(\mathbf{H}')$, the appropriate value of the stationary functional, by back substitution of $\mathbf{H}_f$ into the matrix equations.

A typical problem for which the truly three-dimensional analysis is necessary is calculation of the scattering parameters of a metallic sphere

Fig. 6.12. Finite elements for three-dimensional field analysis.
(a) Hexahedron – 8 nodes and 5 component tetrahedra. (b)
Pentahedron – 6 nodes and 3 component tetrahedra.

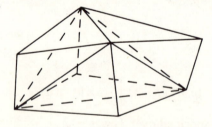

(a)

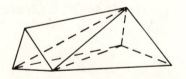

(b)

suspended symmetrically within a rectangular waveguide carrying the TE_{10} fundamental mode. (See Chapter 2, Section 7 for an explanation of modes in a waveguide.) Because of the symmetry, $S_{11} = S_{22}$. Collin (1966) shows, by considering conservation of energy for a lossless arrangement, as here, that if

$$S_{11} = S_{22} = S \exp (j\phi),$$ (4.49)

then

$$S_{12} = \pm j(1 - S^2)^{1/2} \exp (j\phi).$$ (4.50)

Thus, it is only necessary to determine one complex parameter in this case. The **H**-vector working variable was used, so that the curved surface of the sphere and the waveguide walls provided natural (homogeneous Neumann) boundary conditions. There are three planes of symmetry, so that only one-eighth of the problem volume had to be modelled. Nineteen pentahedra and twenty hexahedra, all second-order were employed. The two symmetry planes parallel to the waveguide volume

Fig. 6.13. Discretisation for the metallic sphere problem. One-quarter of the problem is modelled; R = sphere radius; 19 pentahedra and 20 hexahedra were used, all second order. No elements are required within the sphere.

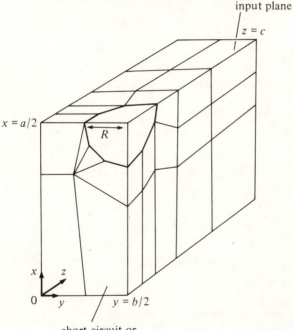

provide homogeneous Dirichlet boundaries (transverse **H** components
to vanish, normal **H** free to vary). The third plane of symmetry, bisecting
the sphere and transverse to the waveguide axis, can be taken either as
a short circuit (Neumann) or open circuit (Dirichlet) plane. This is
obviously equivalent to moving a short circuit plane J_1, well away from
the sphere, into two positions, had the third plane of symmetry not been
exploited. Since only one complex parameter S_{11} is to be determined,
the third position of the shorting plane is not required. The input plane
$z = c$ requires inhomogeneous Dirichlet specification of $\mathbf{H}_t = A(h_x, h_y)$
for the TE_{10} mode, where

$$h_x = \sin(\pi x/a), \qquad h_y = 0 \tag{4.51}$$

(see Ramo, Whinnery & Van Duzer, 1965), and A is an arbitrary
constant. Fig. 6.13 shows the spatial discretisation chosen. The results
(Figs. 6.14 and 6.15) are compared with what appears to be the only
other solution available, that from a perturbation method of Hinken
(1980). There is good agreement for small relative diameter spheres.
However, as might have been expected when comparing results from a

Fig. 6.14. $|S_{\parallel}|$ **versus R for a metallic sphere in a rectangular
waveguide, TE_{10} mode incident.**

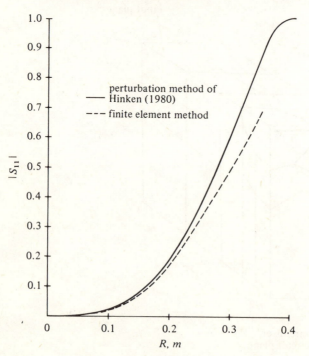

perturbation method, there is disagreement when the sphere diameter approaches the narrow dimension of the waveguide (here the dimensions were 1×2 m). It is to be noticed that the finite element method should not be particularly sensitive as to whether the sphere is small or large, so that it may be claimed that the finite element solution is the more acceptable result where the two methods disagree.

5.   Readings

Three-dimensional finite element analysis for electrical problems is one of the least-developed topics (at the time of writing) to be covered in this book. The lack of present-day coverage of this area represents the difficulty, rather than any unimportance, of the topic. Many of the most pressing problems, magnetostatics analysis of electrical machines, for example, are truly three-dimensional. Historically, the application of finite element techniques to structural mechanics has preceded the use of the method in electrical engineering. Thus some light is thrown on three-dimensional techniques by the standard works

Fig. 6.15. $\angle S_{\parallel}$ **versus radius R for the metallic sphere problem.**

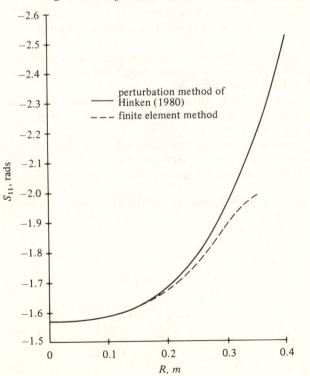

concentrating upon structural problems, such as Zienkiewicz (1977), Norrie & de Vries (1978) and Desai & Abel (1972). However, the different and specialised nature of the problems dealt with there detract from the usefulness to electrical engineers of those writings. The papers by Simkin & Trowbridge (1979) and (1980) give a good background to the very important topic of three-dimensional magnetic field analysis.

References

Collin, R. E. (1966). *Foundation of Microwave Engineering*. New York: McGraw-Hill.

Desai, D. S. & Abel, J. F. (1972). *Introduction to the Finite Element Method*. New York: Van Nostrand.

Ferrari, R. L. & Maile, G. L. (1978). 'Three-dimensional finite element method for solving electromagnetic problems', *Electronics Letters*, vol. 14, no. 15, pp. 467–8.

Hinken, H. H. (1980). 'Conducting spheres in rectangular waveguides', *Institute of Electrical and Electronics Engineers Transactions on Microwave Theory and Techniques*, vol. **MTT–28**, no. 7, pp. 711–14.

Norrie, D. H. & de Vries, G. (1978). *Introduction to Finite Element Analysis*. New York: Academic Press.

Ramo, S., Whinnery, J. R. & Van Duzer, T. (1965). *Fields & Waves in Communications Electronics*. New York: John Wiley.

Riley, K. F. (1974), *Mathematical Methods for the Physical Sciences*. Cambridge University Press.

Simkin, J. & Trowbridge, C. W. (1979). 'On the use of the total scalar potential in the numerical solution of field problems in electromagnetics', *International Journal for Numerical Methods in Engineering*, **14**, 423–40.

Simkin, J. & Trowbridge, C. W. (1980). 'Three-dimensional non-linear electromagnetic field computations, using scalar potentials', *Institution of Electrical Engineers Proceedings*, vol. 127, part B, pp. 368–74.

Silvester, P. (1972). 'Tetrahedral polynomial finite elements for the Helmholtz equation', *International Journal for Numerical Methods in Engineering*, **4**, 405–13.

Sucher, M. & Fox, J. (1963). *Handbook of Microwave Measurements*, vol. 1. New York: Polytechnic Press.

Zienkiewicz, O. C. (1977). *The Finite Element Method*, 3rd edn. London: McGraw-Hill.

7

Numerical solution of finite element equations

1. **Introduction**

In the finite element method, integral or differential equation problems are solved by substituting an equivalent problem of matrix algebra, in the expectation that the techniques of numerical mathematics will allow fairly direct solution of the matrix problem through the use of digital computers. Corresponding to the boundary-value problems or the Fredholm integral equations of electromagnetics analysis, there then results a matrix equation of the form

$$(\mathbf{A} + p\mathbf{B})\mathbf{x} = \mathbf{y}, \qquad (1.01)$$

where $\mathbf{A}$ and $\mathbf{B}$ are matrices, $\mathbf{x}$ is the unknown vector of coefficients, and $\mathbf{y}$ is a known vector, while p is a scalar parameter. This equation has a nontrivial solution if $\mathbf{A}$ is a positive definite matrix and either (1) p is known, and $\mathbf{y}$ does not vanish, or (2) p is unknown, and $\mathbf{y}$ is identically zero. Problems of the former class are usually called deterministic; they produce a single solution vector $\mathbf{x}$ (provided the value of p does not match one of the possible solutions of the second class). The second class are known as eigenvalue problems; they possess as many solutions $\mathbf{x}$ as the matrix has rows, each corresponding to one special value of p. Values of both p and $\mathbf{x}$ are produced by solution methods, so that each solution in fact consists of a scalar–vector pair $(p, \mathbf{x})$. It may appear at first glance that in this case the system of N simultaneous equations yields $N + 1$ answers, the N components of $\mathbf{x}$ and the value of p; however, such is not the case. Since equations of their class are always homogeneous, solutions to eigenvalue problems are unique only up to a multiplicative constant: if $\mathbf{x}$ is a solution, so is $k\mathbf{x}$, where k is any number. Hence the solution vector $\mathbf{x}$ really only contains $N - 1$ independent degrees of freedom.

Many useful two-dimensional field problems produce matrix equations in 100–1000 variables, while three-dimensional vector fields involving complicated geometric structures may require as many as 50 000–250 000 variables. With present-day computers the solution of a hundred simultaneous algebraic equations is a simple matter, regardless of the structure of the coefficient matrix. Thousands of equations can be handled easily, provided the coefficient matrix contains many zeros and the program is so organised as to take advantage of them. With full coefficient matrices, containing few or no zero elements, even 1000 equations is a large number, for the storage of the coefficient matrix alone will require about four megabytes of computer memory!

As a general rule, coefficient matrices derived from integral equations are full or nearly full. On the other hand, discretisation of differential equations by means of finite elements tends to produce sparse matrices because any one nodal variable will be directly connected only to nodal variables which appear in the same finite element. Hence, the number of nonzero entries per matrix row generally depends on the type of element employed, and has little to do with the type of problem. Thus, very large systems of simultaneous equations arising from discretisation of differential equations are usually also very sparse.

This chapter will be devoted to the most common techniques in present-day use for solving deterministic problems arising in finite element analysis. The solution of finite element equations, like much else in the finite element art, is a highly specialised area; the novice is well advised to use clever programs already written by someone else, not to attempt to write his own. The object here is to give the reader some idea of what he may expect to find.

2.   Triangular decomposition

Most present-day finite element analysis programs rely on one version or another of the Gaussian triangular decomposition technique for solving the large systems of simultaneous equations commonly encountered. In its basic form, this method differs very little from that generally learned in school under the name of 'successive elimination': one eliminates one variable at a time until only one equation in one unknown is left. When that one has been determined, the remainder are found by taking the equations in the reverse order, each time substituting the already known values. These two phases of the process are generally called 'forward elimination' and 'back substitution'.

In large-scale finite element programs, the triangular decomposition method is made to operate in the first instance on the coefficient matrix alone. To illustrate, suppose S is a symmetric, positive definite matrix, as finite element matrices typically are. It can be shown that such a matrix may always be written as the product of two triangular matrices,

$$\mathbf{S} = \mathbf{L}\mathbf{L}^T. \tag{2.01}$$

Here, $\mathbf{L}$ is a lower triangular matrix; it has only zero elements above and to the right of its principal diagonal. (The superscript T denotes transposition, so that $\mathbf{L}^T$ is upper triangular.) The process of computing the lower triangular matrix $\mathbf{L}$ is called triangular factorisation, or triangular decomposition; $\mathbf{L}$ is known as the lower triangular factor of $\mathbf{S}$. The requirement that $\mathbf{S}$ must be a positive definite matrix corresponds physically to the requirement that stored energy must be positive for any and all excitations. This requirement is certainly met by physical systems encountered in practice.

Suppose now it is required to solve the matrix equation, which might typically arise from finite element analysis,

$$\mathbf{S}\mathbf{x} = \mathbf{y}. \tag{2.02}$$

To do so, $\mathbf{S}$ is factored as in (2.01), and (2.02) is rewritten as the pair of equations

$$\mathbf{L}\mathbf{z} = \mathbf{y}, \tag{2.03}$$
$$\mathbf{L}^T\mathbf{x} = \mathbf{z}, \tag{2.04}$$

where $\mathbf{z}$ is an auxiliary vector. Because $\mathbf{L}$ is triangular, row k of $\mathbf{L}$ can contain nonzero entries only in its first k columns. In particular, the first row of $\mathbf{L}$ contains only one nonzero element, so that the first component of $\mathbf{z}$ can be computed immediately. The second row of $\mathbf{L}$ contains only two nonzero elements, hence it relates only the first two components of $\mathbf{z}$; but since the first component of $\mathbf{z}$ is known, the second may now be computed. Continuing, the system of equations (2.03) is quickly solved for $\mathbf{z}$. Since $\mathbf{L}^T$ is also triangular, a similar process is used for solving (2.04) to find $\mathbf{x}$; only the sequence of operations is reversed because $\mathbf{L}^T$ is upper rather than lower triangular.

Clearly, the key to equation solving lies in triangular decomposition of the coefficient matrix $\mathbf{S}$. The necessary steps are readily deduced by examining the desired result. Written out in detail, Eq. (2.01) requires that

$$S_{ik} = \sum_{j=1}^{\min(i,k)} L_{ij}L_{kj}. \tag{2.05}$$

Note that the summation only extends to the lower of i or k, for all elements beyond the diagonal of $\mathbf{L}$ must vanish. For the diagonal elements themselves, (2.05) may be solved to yield

$$L_{ii} = \sqrt{\left(S_{ii} - \sum_{j=1}^{i-1} L_{ij}^2\right)}. \tag{2.06}$$

The first row of **L** only contains a diagonal element, which is given by

$$L_{11} = \sqrt{S_{11}}. \tag{2.07}$$

In the second (and every subsequent) row, the summation of Eq. (2.05) contains no more terms than the column number k. Working across each row in the natural sequence, only one new unknown then appears each time. The method is thus easily applied, and may be summarised in the following prescription:

(a) Set

$$L_{11} = \sqrt{S_{11}}. \tag{2.07}$$

(b) In each row i, compute the off-diagonal element in each column k by

$$L_{ik} = \left(S_{ik} - \sum_{j=1}^{k-1} L_{ij} L_{kj} \right) \Big/ L_{kk} \tag{2.08}$$

and the diagonal element by

$$L_{ii} = \sqrt{\left(S_{ii} - \sum_{j=1}^{i-1} L_{ij}^2 \right)}, \tag{2.06}$$

until all of **L** has been calculated.

To illustrate the procedure, consider the symmetric positive definite matrix

$$S = \begin{bmatrix} 4 & 4 & 4 & & & & & \\ 4 & 5 & 2 & & & & & \\ 4 & 2 & 9 & 1 & 2 & 1 & & \\ & & 1 & 10 & -7 & 1 & & \\ & & 2 & -7 & 14 & & 2 & 1 \\ & & 1 & 1 & & 9 & -8 & -2 \\ & & & & 2 & -8 & 9 & \\ & & & & 1 & -2 & & 6 \end{bmatrix}. \tag{2.09}$$

To improve clarity of presentation, all zero entries have been left blank in **S**. By using the technique described, the matrix (2.09) is easily shown to have as its triangular factor **L**,

$$L = \begin{bmatrix} 2 & & & & & & & \\ 2 & 1 & & & & & & \\ 2 & -2 & 1 & & & & & \\ & & 1 & 3 & & & & \\ & & 2 & -3 & 1 & & & \\ & & 1 & & -2 & 2 & & \\ & & & & 2 & -2 & 1 & \\ & & & & 1 & & -2 & 1 \end{bmatrix}. \tag{2.10}$$

Correctness of the decomposition may of course be verified easily by multiplying as indicated in Eq. (2.01).

To complete the example, suppose the right-hand vector **y** to be given by

$$\mathbf{y}^T = [4 \quad 1 \quad 10 \quad -18 \quad 16 \quad 8 \quad -9 \quad 1]. \tag{2.11}$$

Solving (2.03) by the process of forward elimination described above yields the auxiliary vector **z** as

$$\mathbf{z}^T = [2 \quad -3 \quad 0 \quad -6 \quad -2 \quad 2 \quad -1 \quad 1]. \tag{2.12}$$

The solution vector **x** may then be recovered by the process of back substitution, i.e., by solving (2.04):

$$\mathbf{x}^T = [-5 \quad 3 \quad 3 \quad -3 \quad -1 \quad 2 \quad 1 \quad 1]. \tag{2.13}$$

It should be evident that the process of triangular decomposition need only be performed once for any given matrix. If several different right-hand sides are of interest, the forward elimination and back substitution will have to be performed separately for each one. However, as will be shown next, the relative amount of computational work involved in the elimination and back substitution process is not large.

3.   A Choleski decomposition program

To illustrate the techniques employed in equation solving, a simple Choleski decomposition program may prove instructive. Program EQSOLV (p. 202 below) is in fact the program called by the sample finite element program of Chapter 1. It is not optimal in any sense; it does not even exploit matrix symmetry to economise on memory. However, it is probably easy to read and understand.

After the COMMON block which allows for access to all relevant variables in the calling program, subroutine EQSOLV proceeds with triangular decomposition. The Choleski algorithm is used, in exactly the form stated above, with the sole exception that the first two rows (rather than just the first one) is given special treatment. The need to do so arises only from the peculiarities of the Fortran language, in which DO loops may not start with a zero value of the DO index; it has nothing to do with the mathematical method.

The Choleski decomposition method as implemented in EQSOLV fills both upper and lower halves of the matrix storage area with the upper and lower triangular factors. This approach is wasteful of storage, and is not normally used in practical finite element programs. However, the more sophisticated storage arrangements, such as the band or profile methods, render the program indexing more complicated. They are therefore often much more difficult to read and understand.

After the Choleski decomposition has been performed, EQSOLV carries out forward elimination and back substitution. In this process, the right-hand side RTHDSD originally furnished is overwritten with the solution. EQSOLV performs triangular decomposition, forward elimination, and back substitution in a single program, but it is quite usual to subdivide equation-solving programs instead into two or three subroutines. Removing the triangular decomposition to a separate program allows solution with several different right-hand sides, without repeating the decomposition. The penalty paid for doing so is negligible in computer time and space.

4.   Time and storage for decomposition

To assess the efficacy of the triangular decomposition method, some assessment of the total time and memory requirements is necessary. An estimate of both will be given here for dense (fully populated) matrices; refinements useful for sparse matrices will be developed subsequently.

The process of triangular decomposition proceeds step-by-step, computing one new entry of the triangular factor $\mathbf{L}$ at each step. To find element k of row i in $\mathbf{L}$, k arithmetic operations are required, where each 'operation' is the combination of one multiplication or division and one addition or subtraction, along with such integer operations as may be necessary on array subscripts in order to locate the kth element in row i. The number $M(i)$ of operations needed to compute row i of the triangular decompose is the sum of operations required for the elements of that row,

$$M(i) = \sum_{k=1}^{i} k = i(i+1)/2, \tag{4.01}$$

plus one square root operation. The work required for the entire matrix is the sum of the amounts needed for the individual rows, or

$$M = \sum_{k=1}^{N} k(k+1)/2 = N(N+1)(N+2)/6, \tag{4.02}$$

plus N square roots. This amount of work of course covers the decomposition only; nothing is included for actually solving equations. The forward elimination needed for solving with one right-hand side uses up an additional E operations,

$$E = \sum_{k=1}^{N} k = N(N+1)/2. \tag{4.03}$$

Exactly the same number of operations is needed for back substitution.

Thus the amount of work involved in forward elimination and back substitution is, for all practical purposes, equal to the work required for one matrix–vector multiplication, which requires exactly N^2 operations.

Where several right-hand sides are to be solved for with the same coefficient matrix – for example, in calculating the magnetic field in a particular set of coils for several different sets of current values – it may at first glance seem tempting to compute the inverse of the coefficient matrix, and then to multiply each new right-hand side by the inverse. This approach cannot save computing time, and it frequently wastes computer memory, for the following reasons. The most economic procedure for computing the inverse of a matrix is to perform triangular decomposition, and then to solve N times, each right-hand side being one column taken from the unit matrix. While these solutions can be carried out a little more economically than the general case (operations with known zeros can be eliminated by clever programming), the cost of computing an explicit inverse must always be higher than that of triangular decomposition because it actually begins by carrying out a triangular decomposition and then performs a few other operations besides. Subsequent matrix–vector multiplications would entail the same cost as do elimination and back substitution. Hence the invert-and-multiply strategy can never be advantageous as compared to decompose-eliminate-backsubstitute.

The computer memory required to calculate and store the explicit inverse of a matrix can never be less than that needed for the triangular factors, and may often amount to considerably more. Consider again Eq. (2.08), which prescribes the operations required to compute element $L(i, k)$ of the triangular factor $\mathbf{L}$. Suppose element $S(i, 1)$ of the original coefficient matrix $\mathbf{S}$ is zero. Examination of (2.08) shows that the corresponding element of $\mathbf{L}$ is then zero also. Further, if $S(i, 2)$, $S(i, 3)$, $\ldots$, $S(i, k)$ are all zero, then no nonzero terms can appear in the summation in (2.08); the corresponding elements of $\mathbf{L}$ must vanish. Of course, this argument does not hold true for any and all zero elements of $\mathbf{S}$. If there are any nonzero entries to the left of $S(i, k)$, the summation in (2.08) does not in general vanish, and the corresponding entry in $\mathbf{L}$ very likely will not vanish either. Thus one may conclude that the leftmost nonzero element in any given row of $\mathbf{L}$ will in general correspond to the leftmost nonzero element in the same row of $\mathbf{S}$; but from that element up to the diagonal, $\mathbf{L}$ may be more fully populated than $\mathbf{S}$.

Since the leftmost zero elements of the rows of $\mathbf{L}$ correspond exactly to the leftmost zero elements in $\mathbf{S}$, finite element programs are often arranged to avoid storing the leading zeros. Considerable economies of storage may be effected in this way. An explicit inverse of **S**, on the other hand, is full, with no zero entries in the general case. Computing and storing the inverse requires space for a fully populated matrix, while substantially less storage often suffices for the triangular factors. Thus the storage requirement for **L** is bounded above by the storage requirement of the inverse; there cannot occur any case in which computing the inverse can be better from the point of view of storage. Consequently, practical finite element programs never compute and store inverses. They generally either perform triangular decomposition, then eliminate and back substitute; or else they combine the elimination and back substitution with the decomposition so that the triangular factors are not explicitly stored.

5.   Profile and band storage

When triangular decomposition of a matrix is performed, it is usually found that the lower triangular factor **L** is denser (contains fewer zeros) than the original matrix. As noted above, the leftmost nonzero entry in any given row of **L** must in general coincide with the leftmost nonzero entry in the corresponding row of the original matrix **S**. To put the matter another way, any left-edge zeros in **S** are preserved in decomposition. Zeros to the right of the leftmost nonzero element, on the other hand, are not in general preserved; they are said to fill in. Of course, any particular matrix element may fortuitously turn out to be zero because of numerical cancellation. Such zeros are sometimes called 'computed zeros'. The zeros referred to here result from the topological structure of the matrix, and remain zero in any matrix with the same structure, independently of numerical values; they are referred to as 'structural zeros'. By way of example, consider the matrix **S** of Eq. (2.09). The S_{65} and S_{87} entries in this matrix are zero. But comparison with (2.12) shows that they have filled in during the process of decomposition. On the other hand, the S_{64} and S_{86} entries in L are zero, as the result of accidental numerical cancellation. To verify this point, it suffices to decompose another matrix of exactly the same structure as **S**, but with different numerical values. The S_{64} and S_{86} entries of **L** do not in general remain zero, while the left-edge zeros are preserved.

Equation-solving programs generally need to reserve computer memory space for both **S** and **L**. There is clearly no need to reserve space for the left-edge zeros in **S** and **L**, for they are known to remain zero throughout the entire computation. Therefore, it is usual to arrange for storage of matrices in one of several compacted forms. Two of the most common arrangements are band matrix storage and profile storage.

Let **S** be a symmetric matrix. Suppose that the leftmost nonzero entry in row i occurs in column $n(i)$. The half-bandwidth M of **S** is defined as

$$M = 1 + \max_i [i - n(i)],\qquad(5.01)$$

the maximum being taken over all i, $0 < i < N + 1$. Storage in banded form is arranged by storing exactly M numbers for each matrix row. For example, in row k, the first entry stored would be that belonging to column $k - M + 1$, the last one that belonging to column k (the diagonal element). The missing elements in the first few rows are usually simply filled in with zeros. For example, the matrix **S** of Eq. (2.09) can be stored in a single array as

$$\text{band}(\mathbf{S}) = \begin{bmatrix} 0 & 0 & 0 & 4 \\ 0 & 0 & 4 & 5 \\ 0 & 4 & 2 & 9 \\ & & 1 & 10 \\ & 2 & -7 & 14 \\ 1 & 1 & & 9 \\ & 2 & -8 & 9 \\ 1 & -2 & & 6 \end{bmatrix}.\qquad(5.02)$$

In this representation, all zeros contained in the original matrix have been left blank for the sake of clarity. However, also for the sake of clarity, the zeros artificially introduced to fill the leading array elements have been entered explicitly. Even for the quite small matrix of this example, the storage required in banded form is smaller than that needed for full storage, requiring 32 locations (including the artificial zeros) as against 37 (taking advantage of matrix symmetry). For large matrices, the difference can be very large indeed.

The advantage of banded storage is simplicity: only one new item of information is required, the half-bandwidth M. But this form of storage may still be relatively inefficient because many rows are likely to contain at least some structural zeros. Indeed, in the form (5.02) four structural zeros are stored. These are eliminated by the so-called profile storage arrangement. In this arrangement, only those members of each row are stored which fall between the leftmost nonzero entry and the diagonal. The matrix of (2.09) is then stored as a single numeric string:

$$\text{prof}(\mathbf{S}) = [4\ 4\ 5\ 4\ 2\ 9\ 1\ 10\ 2\ -7\ 14\ 1\ 1\ 0\ 9$$
$$2\ -8\ 9\ 1\ -2\ 0\ 6].\qquad(5.03)$$

The proper indices of these values can be recovered only if the position of the leftmost nonzero entry in each row is known. One simple fashion of keeping track of it is to store the leftmost nonzero entry locations $n(i)$, as an integer string:

$$\text{left}(\mathbf{S}) = [1 \ 1 \ 1 \ 3 \ 3 \ 3 \ 5 \ 5]. \tag{5.04}$$

In the case used in this example, the storage requirement for $\mathbf{S}$ is now reduced to 22 locations, plus the array of leftmost locations $n(i)$. As compared to full matrix storage, the reduction is likely to be quite remarkable, in the case of large matrices with a few fully populated or nearly-fully populated rows.

One further method of memory conservation is available to the analyst. Careful reexamination of Eqs. (2.05)–(2.10) will show that once the (i, k) entry of $\mathbf{L}$ has been computed, the corresponding entry of $\mathbf{S}$ is never used again. Consequently, there is no need to store both $\mathbf{S}$ and $\mathbf{L}$: it is sufficient to allocate space for one array only, to deposit $\mathbf{S}$ in this space initially, and then to overwrite $\mathbf{S}$ with $\mathbf{L}$ as decomposition proceeds. Should $\mathbf{S}$ be required for some purpose at a later time, it can be read into memory again from some external storage medium.

The triangular decomposition, forward elimination, and back substitution processes are unaltered in principle when matrices are stored in compact forms. However, the change in data storage implies that the details of subscript computation (i.e., the array indexing in the computer programs) must be modified also. Since there is no need to perform vacuous operations on the left-edge zeros, the summations in Eqs. (2.06) and (2.08) may be altered to run, not from $j = 1$ (the left edge of the full matrix) but from the first nonzero matrix entry. Thus, (2.08) should be replaced by

$$L_{ik} = \left(S_{ik} - \sum_{j=J}^{k-1} L_{ij}L_{kj} \right) \Big/ L_{kk}, \tag{5.05}$$

where the lower limit J of the summation is given by

$$J = i - M \tag{5.06}$$

for band stored matrices because no computations should be executed outside the band edge. For profile stored matrices, no calculation should take place if either factor lies beyond the left edge of its row, so that

$$J = \max[n(i), n(k)]. \tag{5.07}$$

The lower summation limit for Eq. (2.06) of course must be amended correspondingly.

In performing triangular decomposition of a band stored matrix, each off-diagonal element of the lower triangular factor is calculated by a single application of Eq. (5.05), and each diagonal element by

$$L_{ii} = \sqrt{\left(S_{ii} - \sum_{j=J}^{i-1} L_{ij}^2 \right)}. \tag{5.08}$$

It is important to observe that the entries of row i are computed by using entries from the immediately preceding rows, but never from rows numbered lower than $i - M + 1$. Consequently, there is no need to house rows 1, 2, ..., $i - M$ in the immediate-access (core) memory of the computer while work on row i proceeds; only the rows from $i - M + 1$ to i are required. That is to say, the immediate-access memory actually necessary is just that sufficient to house the in-band elements of $M + 1$ rows, or a total of $(M + 1)(M + 2)/2$ numbers. The bandwidth of a sparse matrix is often very much smaller than its total number N of rows. Many large-scale finite element analysis programs therefore use so-called out-of-core banded solvers, which house the matrix $\mathbf{S}$ (in banded storage) on disk, tape, or other external storage media. Triangular decomposition is then performed by computing one or more rows of $\mathbf{L}$, writing them to an output tape or disk, then moving on the remaining rows in immediate-access memory, and reading in one or more additional rows of $\mathbf{S}$. The matrices $\mathbf{S}$ and $\mathbf{L}$ are thus passed through memory, with only $M + 1$ (or fewer) rows ever concurrently resident. Matrices of orders 5000–50 000, with bandwidths of 500 or so, are often treated in this way. A matrix of order 5000 with $M = 500$ requires about 125 000 words of immediate-access memory, a reasonable demand for present-day medium-sized computers, for such out-of-core triangular decomposition and subsequent equation solving. Storage of the matrix itself takes about 2 500 000 words of storage in banded form, or 12 500 000 words if written out without exploitation of sparsity. Thousands of feet of magnetic tape would be required for the latter form of storage.

6. Structure of finite element matrices

Most practical finite element programs exploit matrix sparsity in order to keep both computing time and memory requirements within reasonable bounds. Band matrix and profile methods store all zeros within the band or profile, but reject those outside the band. Keeping the matrix bandwidth small is therefore a matter of prime concern.

Rarely, if ever, do random node numberings on finite elements lead to good matrix sparsity patterns. In most practical programs, nodes are assumed to be randomly numbered at the outset, and some systematic renumbering scheme is then applied to produce an improved sparsity pattern. The only known technique guaranteed to produce the true minimum bandwidth is that of enumerating all possible numberings. Unfortunately, in an N-variable problem there are $N!$ possible numberings, so that the enumerative technique is quite impractical. On the other hand, there do exist several methods which produce nearly-minimal bandwidths or profiles in most problems.

Renumbering methods take no account of the numeric values of matrix entries; they are concerned only with matrix structure. To exhibit structure in a form independent of the numeric values, and without regard to the numbering of variables, it is convenient to give the coefficient matrix a graphical representation. Since every nonzero matrix entry S_{ij} represents a direct connection between variables i and j, one may depict the matrix as a set of linking connections between variables. As an example, Fig. 7.1 shows a structural representation of Eq. (2.09). Here, the variables are shown as circles, while the nonzero matrix entries are indicated by lines linking the variables. For first-order triangular elements, such a graphical representation has exactly the appearance of the finite element mesh itself.

To make the representation of Fig. 7.1 complete, the element values could be entered along each link. They have been omitted, since only the structure of the matrix is of interest, not the numerical values. In fact, the variable numbers could well be omitted, since their numbering is essentially arbitrary anyway.

Computationally, the structure of a matrix may be represented by a set of integer pairs E, $E = (i, j)$, each denoting a pair of variable numbers (i.e., row and column numbers), and each corresponding to a linking line in the graph. The half-bandwidth M of a matrix is then given by the maximal node number difference $i - j$, plus one, taken over all the matrix elements E. That is

$$M = 1 + \max_{\text{all } E} |i - j|. \qquad (6.01)$$

Fig. 7.1. Representation, as an undirected graph, of the symmetric matrix of Eq. (2.09).

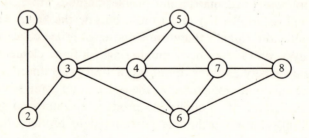

Similarly, the profile (the total storage required for the matrix in profile storage) is given by

$$P = \sum_{\text{all } E} (|i - j| + 1). \tag{6.02}$$

Good node numberings are obviously those which minimise either the bandwidth M or the profile P, depending on which form of storage is contemplated. A good numbering for one purpose is not necessarily good for the other, since a few very large differences $|i - j|$ do not affect the profile P very much, but are ruinous for banded storage. Nevertheless, orderings that produce good bandwidths also give good profiles in a surprising number of practical cases.

Many node numbering methods begin by classifying all the variables into sets called 'levels'. A starting variable, say K, is first chosen and assigned to level $L(0)$. Next, level $L(i)$ is created by taking all the variables which are directly connected to variables of level $L(i-1)$ but which have not been assigned to any of levels $L(0), L(1), \ldots, L(i-1)$. Each and every variable can belong to only one level. For example, starting with variable 1, the levels for the left graph of Fig. 7.2 are

$$\begin{aligned} L(0) &= [1] \\ L(1) &= [2 \quad 3] \\ L(2) &= [4 \quad 5 \quad 6] \\ L(3) &= [7 \quad 8]. \end{aligned} \tag{6.03}$$

There are altogether four levels; in other words, the shortest path between node 1 and the most distant node requires traversing three links. Indeed, it is always possible to reach a node in $L(i)$ by traversing exactly i links from the starting node because $L(i)$ is defined as the set of all nodes directly connected (one step away from) nodes in $L(i-1)$.

The number of levels is dependent on the choice of starting node. For the left half of Fig. 7.2, but this time starting with variable 3, the

Fig. 7.2. Two possible node (variable) numberings, which do not yield the same matrix bandwidth.

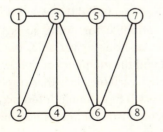 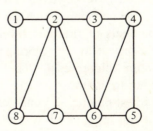

level structure is quickly obtained as

$$L(0) = [3]$$
$$L(1) = [1 \ 2 \ 4 \ 5 \ 6] \tag{6.04}$$
$$L(2) = [7 \ 8].$$

The total number of levels, as well as the grouping of variables within the levels, depends on the starting node. If all possible starting nodes are tried, and the highest-numbered level reached for any of them is $L(K)$, the matrix is said to have a topological diameter of K.

7.   Numbering techniques

There exists a substantial number of specific renumbering techniques for exploiting sparsity. No practical method known to date is guaranteed to produce the truly optimal numbering, but several come fairly close in most cases. This area is still developing, so it is not appropriate to give a detailed treatment of the available methods here; but since all the available methods are based essentially on examination of a graphical representation of matrix structure, certain common principles can be stated.

For the left graph of Fig. 7.2, the numbering on the left is superior to that on the right because the variable number differences $|i-j|$ are smaller. Correspondingly, the nonzero matrix elements cluster more tightly around the principal diagonal. Good numberings are usually such that nodes with widely different numbers are widely separated. Such separation is obtained if all nodes in level $L(i)$ are numbered before those in level $L(i+1)$. In other words, once the starting node has been assigned the number 1, the members of level $L(1)$ should be numbered next, then those of $L(2), \ldots$, until no more levels are left. While the general principle is simple enough, it leaves open the questions of how to choose the starting node, and how to number the members of each level. One quite effective technique chooses the node with the fewest connections to all others as the starting node. A level structure is then constructed, starting from that node. Next, within each level one numbers first those members which are directly connected to the smallest number of members of the next level. The rationale for doing so is that the longer one delays numbering a particular node or variable, the smaller the index differences $|i-j|$ between it and the members of the next level; numbering the most strongly connected variables last thus implies that the largest number of connecting links will have small differences $|i-j|$. While this argument is essentially probabilistic and gives no guarantees as to quality of numbering, it does work quite well in a large number of cases.

To take the case of Fig. 7.2 again, with the level structure as in (6.03), the nodes have the following number of connections to nodes of the next higher level:

node:	1	2	3	4	5	6	7	8
level:	0	1	1	2	2	2	3	3
connections:	2	1	3	0	1	2	0	0

It is readily verified that the numbering on the left of Fig. 7.2 is in fact the numbering given by the above method.

From (6.03) and (6.04) it should be clear that the best numberings are likely to result from those choices of starting point which yield many levels, each containing only a few nodes. This observation leads to another group of methods, in which the matrix structure is examined to find its topological diameter, and the starting node is chosen at one end of a true diameter. While the amount of work required to find the new variable numbering is generally increased by the extra operations needed to find the diameter, the additional effort is usually amply repaid by decreased costs of actually solving equations afterwards.

8.   Readings

There are many good textbooks on the solution of simultaneous equations, to which the reader may refer with profit. However, finite element matrices have many peculiar properties so that these quite general works may not suffice. Thus, finite element textbooks in the structural and mechanical engineering areas may be consulted; although the physical problems there quite often differ profoundly from those of electromagnetics, the structure of the finite element equations is frequently the same. The book by Bathe & Wilson (1980) is specifically concerned with numerical methods applicable to finite elements, and is therefore to be recommended. Irons (1979) is less mathematically oriented but evaluates techniques and suggests ways of avoiding pitfalls based on practical experience.

The various renumbering methods now available are less well documented in textbook literature, so that recourse must be had to the periodical literature. The first, and still popular, method is that of Cuthill & McKee (1969), which was superseded in due course by the similar, but more effective, algorithm of Gibbs, Poole, and Stockmeyer (1976). A very fast method, which has not yet been extensively implemented, is that described by George & Liu (1980).

References

Bathe, K.-J. & Wilson, E. (1980). Numerical Methods in Finite Element Analysis. New York: McGraw-Hill.

Cuthill, E. & McKee, J. (1969). 'Reducing the bandwidth of sparse symmetric matrices', *Proceedings of the 24th National Conference of the Association for Computing Machinery* (ACM Publication P–69), pp. 157–72. New York: Association for Computing Machinery.

George, A. & Liu, J. W. H. (1980). 'A minimal storage implementation of the minimum degree algorithm', *Society for Industrial and Applied Mathematics Journal on Numerical Analysis*, vol. 17, no. 2, pp. 282–99.

Gibbs, N. E., Poole, W. G., Jr. & Stockmeyer, P. K. (1976). 'An algorithm for reducing bandwidth and profile of a sparse matrix', *Society for Industrial and Applied Mathematics Journal on Numerical Analysis*, vol. 13, no. 2, pp. 236–50.

Irons, B. & Ahmad, S. (1979). *Techniques of Finite Elements*. Chichester: Ellis Horwood.

The EQSOLV equation solver program

```
C
C******************************************************************
C
      SUBROUTINE EQSOLV
C
******************************************************************
C
C     SOLVES THE SYSTEM OF EQUATIONS WITH SQUARE, SYMMETRIC COEF-
C     FICIENT MATRIX S,
C
C         S * POTENT  =  RTHDSD
C
C     THE PROCEDURE IS CHOLESKY DECOMPOSITION FOLLOWED BY FORWARD
C     ELIMINATION AND BACK SUBSTITUTION.  S AND RTHDSD  ARE  NOT
C     PRESERVED.
C
C================================================================
C     DATA ACCESS COMMON BLOCK -- SAME IN ALL PROGRAM SEGMENTS
C
      LOGICAL CONSTR
      DIMENSION X(50), Y(50), S(50,50)
      DIMENSION CONSTR(50), RTHDSD(50), POTENT(50)
      DIMENSION NVTX(3,75), SOURCE(75), SELM(3,3), TELM(3,3),
     1          INTG(3)
      COMMON INPUT, KONSOL, IERR, NVE, MAXNOD, MAXELM, NODES,
     1       NELMTS, X, Y, NVTX, SOURCE, CONSTR, POTENT, RTHDSD,
     2       SELM, TELM, S, INTG
C================================================================
C
C     CHOLESKY DECOMPOSITION:  REPLACE S BY ITS UPPER AND LOWER
C     TRIANGULAR FACTORS.
C
      S(1,1) = SQRT(S(1,1))
      S(2,1) = S(2,1) / S(1,1)
      S(1,2) = S(2,1)
      S(2,2) = SQRT(S(2,2) - S(2,1)**2)
```

```
C
            DO 60 IROW = 3, NODES
            IROW1 = IROW - 1
            S(IROW, 1) = S(IROW, 1) / S(1, 1)
            S(1, IROW) = S(IROW, 1)
                DO 40 ICOL = 2, IROW1
                ICOL1 = ICOL - 1
                SUM = 0.
                    DO 30 K = 1, ICOL1
                    SUM = SUM + S(IROW, K) * S(ICOL, K)
      30            CONTINUE
                S(IROW, ICOL) = (S(IROW, ICOL) - SUM) / S(ICOL, ICOL)
                S(ICOL, IROW) = S(IROW, ICOL)
      40        CONTINUE
            SUM = 0.
                DO 50 K = 1, IROW1
      50        SUM = SUM + S(IROW, K) * S(IROW, K)
            S(IROW, IROW) = SQRT(S(IROW, IROW) - SUM)
      60    CONTINUE
C
C     FORWARD ELIMINATE.   RTHDSD IS DESTROYED IN THE PROCESS.
C
      RTHDSD(1) = RTHDSD(1) / S(1, 1)
C
            DO 130 IROW = 2, NODES
            IROW1 = IROW - 1
            SUM = 0.
                DO 110 ICOL = 1, IROW1
     110        SUM = SUM + S(IROW, ICOL) * RTHDSD(ICOL)
            RTHDSD(IROW) = (RTHDSD(IROW) - SUM) / S(IROW, IROW)
     130    CONTINUE
C
C     BACK SUBSTITUTE, USING UPPER TRIANGULAR FACTOR
C
      POTENT(NODES) = RTHDSD(NODES) / S(NODES, NODES)
            DO 230 K = 2, NODES
            IROW = NODES + 1 - K
            SUM = 0.
            IROW1 = IROW + 1
                DO 220 ICOL = IROW1, NODES
     220        SUM = SUM + S(IROW, ICOL) * POTENT(ICOL)
            POTENT(IROW) = (RTHDSD(IROW) - SUM) / S(IROW, IROW)
     230    CONTINUE
C
     300 RETURN
         END
```

APPENDIX
Calculations on triangular elements

The matrix element calculations given in Chapter 3, and the corresponding three-dimensional work of Chapter 6, rely on two mathematical facts: the integration formula given in Eq. (5.12) of Chapter 3, and the trigonometric identity of Eq. (5.07) of the same chapter. Since the reader may be interested in how these key facts were arrived at, a short derivation is given here.

1.   Integration in homogeneous coordinates

In computing element matrices for triangular elements, it is necessary to evaluate the definite integral $I(i, j, k)$,

$$I(i, j, k) = \int \zeta_1^i \zeta_2^j \zeta_3^k \frac{d\Omega}{\Omega}. \tag{1.01}$$

This task is best accomplished by noting that in any area integration, the area element may be written in whatever coordinates may be convenient, provided the Jacobian of the coordinate transformation is included:

$$d\zeta_1 \, d\zeta_2 = \frac{\partial(\zeta_1, \zeta_2)}{\partial(x, y)} \, dx \, dy. \tag{1.02}$$

Since the transformation between Cartesian and triangle coordinates is given by

$$\zeta_1 = (a_1 + b_1 x + c_1 y)/(2A), \tag{1.03}$$
$$\zeta_2 = (a_2 + b_2 x + c_2 y)/(2A), \tag{1.04}$$

where A is the triangle area, the Jacobian is readily evaluated:

$$\frac{\partial(\zeta_1, \zeta_2)}{\partial(x, y)} = \frac{1}{2A} \begin{vmatrix} b_1 & c_1 \\ b_2 & c_2 \end{vmatrix} = \frac{1}{2A}. \tag{1.05}$$

The integral $I(i, j, k)$ may therefore be written in the form of an iterated integral

$$I(i, j, k) = 2 \int_0^1 \int_0^{1-\zeta_1} \zeta_1^i \zeta_2^j (1 - \zeta_1 - \zeta_2)^k \, d\zeta_2 \, d\zeta_1 . \tag{1.06}$$

Integrating by parts, one obtains

$$I(i, j, k) = \frac{2k}{j+1} \int_0^1 \int_0^{1-\zeta_1} \zeta_1^i \zeta_2^{j+1} (1 - \zeta_1 - \zeta_2)^{k-1} \, d\zeta_2 \, d\zeta_1 \tag{1.07}$$

and hence

$$I(i, j, k) = [k/(j+1)] I(i, j+1, k-1) . \tag{1.08}$$

Applying (1.08) repeatedly, there then results

$$I(i, j, k) = \frac{j! \, k!}{(j+k)!} I(i, j+k, 0) \tag{1.09}$$

and

$$I(i, j, k) = \frac{i! \, j! \, k!}{(i+j+k)!} I(0, i+j+k, 0) \tag{1.10}$$

But only one of the triangle coordinates actually appears in the integral $I(0, i+j+k, 0)$ on the right of (1.10). Evaluation is therefore straightforward, and yields

$$I(i, j, k) = \frac{i! \, j! \, k! \, 2!}{(i+j+k+2)!}, \tag{1.11}$$

as given in Chapter 3.

Fig. A.1. Geometry of an arbitrary triangle, for proof of the cotangent identity of Chapter 3.

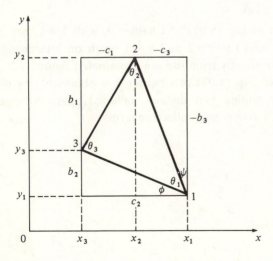

A similar method is applicable to tetrahedra, as indicated in Chapter 6; the main difference is simply that integration is carried out over four homogeneous coordinates rather than three.

2.   The cotangent identity

Equation (5.07) of Chapter 3 is readily derived from basic trigonometric considerations. Let a triangle be given, and let it be circumscribed by a rectangle whose sides are aligned with the Cartesian axes, as in Fig. A.1. The three vertices of the triangle subdivide the rectangle sides into segments given by the numbers

$$b_i = y_{i+1} - y_{i-1}, \tag{2.01}$$

$$c_i = x_{i-1} - x_{i+1}. \tag{2.02}$$

These are the same as Eqs. (5.05)–(5.06) of Chapter 3. From Fig. A.1, it is evident that

$$\cot \theta_1 = \cot \left(\frac{\pi}{2} - \varphi - \psi \right), \tag{2.03}$$

which may be written

$$\cot \theta_1 = \frac{\tan \varphi + \tan \psi}{1 - \tan \varphi \tan \psi} \tag{2.04}$$

or, substituting the ratios of lengths for the tangents of included angles,

$$\cot \theta_1 = - \frac{b_2 b_3 + c_2 c_3}{c_3 b_2 - b_3 c_2}. \tag{2.05}$$

The numerator may now be recognised as being twice the triangle area,

$$\cot \theta_1 = - \frac{b_2 b_3 + c_2 c_3}{2A} \tag{2.06}$$

so that the first part of Eq. (5.07) of Chapter 3, with $k = 1$, has been obtained. Similar results for $k = 2$ and $k = 3$ result on interchange of vertex numbering, or directly from the same geometric figure.

The second part of Eq. (5.07) of Chapter 3 is obtainable by direct algebraic means by adding two distinct included angle cotangents, expressed as in (2.05) above, and collecting terms.

Index